Feeding Pasture-Raised Poultry

By Jeff Mattocks
Poultry and Livestock Nutritionist

The Fertrell Company
Bainbridge, PA
800-347-1566
jeffmattocks@fertrell.com

Feeding Pasture-Raised Poultry
Second Edition

Copyright 2013
by The Fertrell Company
PO Box 265
Bainbridge, PA 17502
www.fertrell.com

Jody Padgham, Editor and Designer

*This book is dedicated to Jack Robinette,
for his decade of mentoring and guidance.*

Thanks to our reviewers:
Susan Beal
Seth J. Epler

Thanks also to the many farmers who have, over the years, contributed to the information collected in this book.

All rights reserved. No part of this book may be reproduced, stored in a retrieval system or transmitted in any form by any means - electronic, mechanical, photocopying or other - without written permission from the Fertrell Company.

The information in this book is presented by the author and believed to be true and complete. All recommendations are made without guarantee on the part of the Fertrell Company. The editor and publisher disclaim any liability in connection with the use of this information.

Printed in the United States
ISBN 978-0-9893328-0-4

About the Fertrell Company

The Fertrell Company is a private company in Bainbridge, Pennsylvania specializing in natural agricultural products and services. It was founded in 1946 by John Johnson of Fullerton, Maryland, a rose enthusiast seeking better rose blooms. Blending rock powder, animal offal and tankage, John developed a superior rose plant food. Thanks to friends and neighbors encouraging him to sell his fertilizer, the Natural Development Company was born. In later years John met Rufus Miller, a Lancaster county, Pennsylvania, native, who in 1964 convinced John to relocate to property he owned in Bainbridge, Pa. A manufacturing facility was built on the bottom three acres of the Miller farm, where Fertrell still resides today. In the mid 1970s Rufus began developing supplements for livestock, starting with dairy minerals to support the Lancaster county dairy industry. Products were soon developed for beef, swine, and eventually poultry. Learn more at www.fertrell.com.

Table of Contents

Page

Introduction..7

The Basics... 8

Feeding Considerations
Rate of Gain.. 10
Feeding Regimes.. 12
Appropriate Water and Feeder Space and Position............... 16
Disease-Causing Organisms...................................... 17
Water Quality... 18

True Grit: The Lowdown on an Important Avian Supplement..21

Feed Sources and Preparation
Commercial Feeds... 24
Milling.. 26
Soybean Roasting... 27
Feed Storage.. 28

Feed Quality Issues..29

Evaluating Your Feed Cost... 31

Poultry Feed Ingredients
Common Ingredients.. 34
Less Common Ingredients.. 41
Vegetarian Diets vs. Meat Diets............................... 44
Soy Alternatives... 44

Fundamental Ration Formulation
Feed Nutrient Values.. 46
Poultry Nutritional Requirements............................. 46
Sample Commercial Rations......................................47

Sample Rations for Pastured Poultry.. 48

Alternative Rations..58

Formulating Your Own Rations..60

Feeding Ducks... 61

Multi-Species Ration..63

Laying Hen Considerations... 66

Diagnose and Treat Common Illnesses................................... 68

Continued

Feeding Pasture-Raised Poultry

Table of Contents *continued*

Page

Appendices
- A: Poultry Nutri-Balancer Study.. 73
- B: No-Soy Ration Research... 75
- C: Feed Ingredient Values and Ration Calculator................. 90
- D: Commercial Broiler-Roaster Nutritional Requirements. 95
- E: Commercial Layer Nutritional Requirements..................... 96
- F: Commercial Turkey Nutritional Requirements.................. 97
- G: Commercial Meat Duck Nutritional Requirements.......... 98
- H: Commercial Broiler Sample Rations..................................... 99
- I: Commercial Roaster Sample Rations................................... 100
- J: Commercial Layer Sample Rations......................................101
- K: Commercial Turkey Sample Rations...................................102
- L: Formulations of Rations Using Pearson Square................103

Endnotes... 106

Additional Resources... 107

Index... 108

Introduction

People have been raising poultry for food for many centuries. Less than 100 years ago throughout America most rural households managed both a small layer flock and a meat flock.

Before the 1920s chickens were not considered a viable livestock because they did not do well in the winter. Thus, eggs were considered a luxury food. Starting in the early 1920s producers recognized that congregating poultry in large production barns and using ever more specialized genetic strains created efficiencies, and thus lowered the costs of production. The trend toward large flocks of closely-contained, genetically-homogeneous birds led to the commodification of the poultry industry. The diets and health issues of the birds involved in this industry have been closely studied and highly refined.

In the past 20 years, there has been a movement among producers to 'reclaim' the fine quality and taste of poultry that has been raised in smaller flocks, or batches, and allowed outside on pasture. Whether one is using the same genetic stock as that developed for the large confinement industry, older heirloom breeds, or newer breeds developed specifically for the growing pastured poultry industry, the nutritional needs and health issues are slightly different from those of large commodity flocks.

It is pretty straight-forward to raise poultry. However, those wishing to be cost-effective and efficient will gain a lot by paying attention to the finer details of their poultry's nutritional needs. The feed ingredients, feed freshness and quality, amount of feed fed, types of nutritional supplements, feeding equipment, water quality and watering equipment all affect the birds' health–and the cost to raise them. Those raising poultry to make a profit will want to pay attention to these issues in order to be cost-effective in their production. Those who don't care about profit will still benefit from the concepts presented here through the maintenance of bird health and productivity.

This book has been developed to help those raising poultry on pasture to understand their options for meeting the nutritional needs of their birds; whether broilers, laying hens, turkeys, or ducks. The recommendations offered here have been based on scientific research, but also reflect the author's many years of experience–working directly with successful pastured poultry producers.

Feeding Pasture-Raised Poultry

The Basics

There are some basic conditions required for any young birds, be they day-old chicks, ducklings or turkey poults.

In the Brooder

• Success with any kind of poultry begins in the brooder. Brooder conditions are critical to the long-term health and survivability of the birds. Cleanliness, good air quality and dry conditions are essential.
• Start at 90° F temperature for the first week, then drop one degree every 1 to 2 days reaching 75° F by week 3. Being too hot is as bad as being too cold.
• Round all interior corners to prevent chick pile up and smothering.
• Bedding preferences in order: peat moss, pine shavings, kiln-dried sawdust, oat hulls, chopped or ground straw, ground corn cobs.

Living Space

• Plan on ½ square foot of space per chick, and increase this to one square foot per bird by the time they go to pasture. In pasture, if broilers are confined in a pen, plan for 1.5 square foot of pen space per bird by 5 weeks of age.
• Mature laying hens should have a minimum of 4 square foot per bird per day when on pasture.
• Turkeys will need as much as 6 square feet per bird while on pasture.
• These spaces are to develop proper social skills, i.e. "pecking order."
• It is standard practice to move broilers to pasture at 3 weeks, although this can vary depending on weather conditions.
• Turkeys (poults) and laying hens will need longer time in the brooder–as much as 5 to 8 weeks. This depends on outdoor climate and poult development. Poults and young layers should be well established on feed, fully feathered and not dependant on supplemental heat sources before they are moved outside.

Feeding

• Start in the brooder with one inch of feeder space per chick, increase by ¼ inch/bird each week as they grow, to a total of 3 to 4 inches/bird by the time the birds are 8 weeks old.

The Basics

• The above recommendation is far greater than the industry standards. But, the industry uses automation to feed 8 to 10 times per day. A pasture poultry producer tends to feed once, maybe twice per day. Fewer feeding times leads to greater competition for space at the feeder. Optimum feeder space will preclude many health and performance issues:
    o More uniform growth and development—fewer runts!
    o Less skin scratched from toe nails, which can lead to Infectious Process (IP), where an open wound allows bacteria such as *E. coli* to enter and cause infections. These infections generally lead to a carcass being condemned for human consumption.
• The ideal goal is to grow Cornish cross broilers on pasture to 4 pounds processed weights at 6 ½ weeks of age (45 days). When properly managed, this can be achieved with less than 12 pounds of feed per bird.
    o Most modern Cornish cross strains are genetically selected to grow out in 35 days for a 3 to 3.5 pound carcass weight. Thanks, fast food fried chicken!
    o Other genetics have an expected grow out of 45 to 70 days, variable depending on the desired finished weight of the bird.
• Keep feed fresh. The starch molecules in the feed will start to oxidize immediately after processing. At 14 days past grinding the feed will begin to taste like stale bread to the poultry. They will eat it, but not as vigorously. Try not to feed food that is 30 days post-grinding or processing. The poultry will consume fresh feed much more vigorously and convert the calories more efficiently.

Water

• Start in the brooder with one inch of waterer space per chick, increase by ¼ inch/bird weekly as they grow, to a total of 3 to 4 inches/bird by the time they are 8 weeks old.
• Birds will consume about twice as much water as feed (by weight not volume) when the water is 75° F. As the air temperature goes up, the water-to-feed ratio goes up. If the birds are consuming more water than this it is an indication that there may be health issues.
• Diseases like coccidiosis and necrotic enteritis due to stomach erosion will cause the birds to consume more water.
• Ideal water temperature is 75° to 80° F.
• Ducks will want additional water. They will play with the water, however, so it is important to either put the water in a setup where the bedding won't get wet, or have the water troughs high enough so ducklings can only drink and not play.

Life Cycle

• Cornish cross broilers should reach 4 pound carcass weights in 6 ½ to 7 weeks.

Feeding Pasture-Raised Poultry

• Heirloom chickens will be much slower, with variation by breed. Plan for at least 12 weeks of grow out for a 4 pound carcass, and on feeding 5 pounds of feed per pound of finished carcass.
• White Broad Breasted turkeys are generally grown for 4 to 5 months to produce 15 to 25 lb carcasses. Assume 3 to 4 pounds of feed per pound of finished carcass for commercial strains of turkeys. I allow 60 pounds of feed per turkey from day one to finish. As turkeys are far better foragers than chickens, this can vary by up to 10% depending on their range and forage base.
• Production breed laying hens will generally start laying eggs at 20 weeks of age, with peak productivity at 30 weeks. See details in "Laying Hen Considerations."
• Meat ducks are most frequently butchered at 8 weeks of age, and will consume 35 to 40 pounds of feed to reach a 4 to 5 pound carcass.

Feeding Considerations

Rate of Gain

Birds will be fed different diets, using various ration formulations, depending on production goals. One of the major determining factors on ration ingredients and balance will be the rate of gain desired for the poultry. This will vary by the species of bird and the production style being used.

• **Fast rate of gain:** *2.1 to 2.3 pounds of feed per pound of live weight, yielding 3.75 pound saleable carcass in 35 days.* These are diets typically fed to meat production poultry to maximize growth potential, used in a commercial environment with climate control devices, lighting control and automated feeding equipment. These feeding methods may be used by pastured poultry producers with the understanding that birds may experience higher mortality rates than non-pasture operations, due to the uncontrolled outdoor environment.

Cornish cross broilers and roasters are examples of fast rate of gain poultry. Broilers are meat chickens raised for 35 to 42 days with a desired weight of 2.75 to 3.5 lb dressed. Broilers are typically fed diets categorized as fast rate of gain. Roasters are meat chickens raised for 49 to 56 days with a desired weight of 4 to 6 lb dressed. Roasters are typically fed diets categorized as slower rate of gain for the last 3 weeks of life.

• **Intermediate rate of gain**: *4 to 5 pounds of grain per pound of live weight, yielding 3.5 to 4.5 pounds of salable carcass in 70 to 84 days.* These feeding programs are used for heritage heavy breeds, Freedom Rangers, Label Rouge, and other slower-growing meat or dual purpose poultry. Cornish cross producers who target a 4 to 5 pound salable carcass at 56 days may also choose these methods.

• **Slow rate of gain:** *5 to 6 pounds of feed per pound of live weight, yielding 3.5 to 4.5 pounds of carcass in 84 to 98 days.* These types of feeding programs are used for egg production for both common layers and breeder stock poultry. Heritage breed poultry also require slower rate of gain feeding systems, as their metabolism has not been bred for rapid development. This feeding style is intentionally designed to allow the poultry to develop a stronger metabolism and immune system and to control the fat production of the poultry. Excess fat will create internal fat build up on organs, primarily those of the reproductive system, causing decreased rate of lay and increased egg size.

Feeding Pasture-Raised Poultry

Turkeys, depending on the genetic potential, can be fed in any of the previously mentioned methods, fast or slow rates of gain.

• **Faster rate of gain turkeys:** This feeding method is for commercial heavy broad breasted breeds, White and Bronze. These breed types have the genetic potential to convert feed to salable carcass at the ratio of 3 to 1 in optimum conditions. They should reach 12 to 25 pound carcass weight in 15 to 20 weeks. Hens at 15 weeks old will produce a 12 pound carcass, toms at 20 weeks old will produce 25 pound carcasses.

• **Slower rate of gain turkeys:** This feeding method will be best suited for heritage breed turkeys, i.e. Bourbon Reds, Narragansetts, Blue Slates, etc. Feeding turkeys for 28 to 32 weeks will produce carcass yields of 12 to 22 pounds. The slower growing breeds typically consume 5 to 6 pounds of feed per pound of saleable (finished) carcass weight.

Feeding Regimes

Restricted vs. Full Feeding

Restricted feeding methods are commonly used in commercial broiler and breeder operations. Broiler feed is restricted to reduce late-term mortality and ascites (water belly). Experiments in which feed is restricted or withheld from broilers from day 7 through day 28–limiting feed to 8 hours only–have found a significant reduction of ascites and late-term mortality (heart attack) rates. Adjusting to lower energy values allows internal organs to develop commensurate with body growth rate, controlling ascites. This type of feeding program is generally accomplished with mash-type diets.

Full feed methods of feeding (allowing the birds to eat as much as they want, at any time of the day) can be used for the development of broilers, pullets and layers when diets are balanced for this type of feeding (as explained in upcoming chapters). This feeding method is not suitable for breeder flocks. The natural tendencies of breeder layers will cause them to over-eat, and thus over-develop, if they are allowed full feed. Feed restriction for broiler breeders is for weight and development control.

Self-Selection and Whole Grain Feeding

Poultry are capable of diet self-selection and eating whole grain feeds. This is nothing new, as evidenced by birds in the wild eating whole seeds at backyard bird feeders. The concept of feeding whole grains to production birds has been swept to the wayside with technology. However, in recent years, the concept of whole grain feeding has resurfaced.

Feeding Considerations

"As with most other classes of poultry, the turkey seems able to balance its own nutrient intake when offered a selection of different diets on a range of individual ingredients. However, results from studies with turkeys are encouraging in that leaner carcasses are produced with this type of system." [1]

"Chickens that are given the opportunity to simultaneously consume two or more feeds that differ significantly in protein content will tend to consume a mix of feeds that is close to the optimum protein content for their stage of growth." [2]

In my work with growers throughout North America, I have found these reflections to be true. Fertrell has managed whole grain field trials with turkey producers. Starting at week 8 we introduce whole wheat along with the regularly prepared feed. In week 8 the consumption of wheat is approximately 10% of the diet. Each progressing week the ratio of wheat and prepared feed gets closer, and will eventually invert as the birds select wheat in larger proportion than prepared feed. By the last week the turkeys will consume approximately 80% whole wheat to 20% prepared feed.

"Feeding whole grains along with pellet-processed feed could result in considerable feed cost savings, depending upon the production system and market conditions. Moreover, some (bird) health benefits could be realized if a proper portion of the bird's diet contained whole grains." [3]

"The feeding of whole grain to poultry was a common practice at one time...In addition to reduced feed cost, there are other good reasons for feeding whole grain. Whole grain frequently has considerable microbial activity on its surface... The feeding of whole grain usually results in a reduction in water consumption and improved litter quality. With improved litter quality comes reduced leg problems and lower coccidial load." [4]

Keep in mind that the poultry MUST have an adequate supply of good quality grit at all times in order to digest whole grains. This is important not only when whole grains are fed, but with mash and foraged feeds as well. (See page 21).

Most of the trials Fertrell performed used wheat as the whole grain. Other grains may be tried. For instance, ducks and geese prefer cracked or whole corn. Turkeys tend to ignore corn and prefer whole wheat. Chickens will generally prefer whole corn. Feeding whole corn to chickens or any other form of poultry should only be done when they are adult, or nearing adult stage of life.

Omega-3 Fatty Acids

Consumers are drawn to Omega-3s due to reports that the acids help in fighting cancer and maintaining health. Higher Omega-3 fatty acid levels in eggs and meat can be achieved by adding feed ingredients such as listed on page 14.

Feeding Pasture-Raised Poultry

Ingredient	Omega-3 Fatty Acid Content [5]
Fish oil (menhaden)	34.7%
Linseed oil	56.4%
Flax seed	5.3%
Canola	4%
Fishmeal (menhaden)	1.75%

These Omega-3 fatty acid sources are all acceptable feed ingredients, but each can create unfavorable side effects when overfed or mixed with others on the list. Too much fish oil or fishmeal will pass on a fishy flavor and smell. Canola will also pass on fishy flavors and aromas. This can occur when these ingredients are combined or when used as lone ingredients added to a ration.

Fishmeal can be added up to 5% of the ration without any off-flavoring side effects, but no higher. Most feed-grade fish meals contain 10% fish oil. If you have access to a higher oil content fishmeal you should calculate a maximum amount that can be added so as to not exceed 5% of the total ration (or 10 lbs of oil per ton of mixed feed). Canola can be added up to 10% of the total ration without noticeable smell or flavor changes.

Flax products, either the seed or the oil, will also create off-smells and flavors. These products transfer a paint-like smell to the eggs or meats produced. The upper limit for flax seeds in a poultry diet is 15%. Flax seed oil should not be used above 1% of the total mix. Additions of flax seed oil above 1% of the total mix should be approached cautiously and gradually.

"Feeding flax seeds to poultry results in direct incorporation of linolenic acid into poultry meat and also into eggs. Feeding laying hens 10% flax results in a 10 fold increase in egg linolenic acid content, while feeding 20 and 30% results in 23 and 39 fold increase respectively... Linolenic acid is essentially responsible for the characteristic smell of fish oils and undoubtedly flax oil does have a paint-type smell. There is some concern about the taste and smell of linolenic acid-enriched eggs, and this area needs more careful study with controlled taste panel work."[6]

Many commercial egg producers have adopted this concept and are adding flax seed to commercial layer diets to increase Omega-3 content of the eggs produced. This can be an excellent marketing tool, as it is now recognized by consumers.

Pastured poultry producers raise chickens on grass to achieve higher levels of Omega-3 fatty acids. I am not aware of the Omega-3 content of grass, however it is my feeling that the precursors and required nutrients are in the green forages and that the animal that ingests the forage manufactures Omega-3s through digestion and enzyme activity. However, although this has been proven in ruminants, it is unknown in poultry, and currently is just a theory.

Feeding Considerations

Organic Poultry Feeding

There is quite a stir in the market place, with consumers looking for products produced in a different way than "conventional" food. This has created a growing market for certified organic poultry—both meat and eggs. To raise organic poultry, one must feed 100% certified organic feed and raise the birds on certified organic pasture. Antibiotics, growth stimulants and other prohibited synthetic materials cannot be fed to the birds at any time during their life.

Due to high demands for all kinds of organic meat in the past few years, the supply of organic grain for feed has gotten tight and the price has increased. Organic grain is raised without the use of chemical fertilizers, herbicides, pesticides, fungicides and genetically modified organisms (GMOs). Any feed ingredient derived from an agricultural commodity must be from a certified organic source. (i.e alfalfa meal, wheat middlings, rice hulls, etc.). Additional restrictions regulate feed supplements. All feed ingredients, supplements and additives must be reviewed for use before they can be used in a certified organic production system.

Fishmeal that is preserved with ethoxyquin is prohibited. Methionine, necessary for proper feather and egg production, is the only synthetic amino acid allowed in organic production, and is regulated to minimal usage levels in organic poultry production. It is anticipated that the allowance of methionine will be phased out in the future.

Organic producers undergo very thorough screening and must adhere to guidelines to maintain organic certification. Organic requirements are tightly enforced and the additional paperwork will take time to set up and get used to. In the first year it may take several hours to gather all of the necessary documentation for organic certification. After the initial certification, each renewal year becomes easier. Someone who keeps good records and is organized should find that paperwork preparation for organic renewal will take only a few hours annually.

Organic certification is under the control of the USDA National Organic Program. Additional information may be found at www.ams.usda.gov/nop. The national rules are governed and enforced through local organizations (certification agencies) that have applied to the USDA to perform certification review. These accredited certification agencies (ACA) will have lists of materials approved for use in organic operations. It is imperative that before you use any new product you verify its compatibility with organic production, regardless of the manufacturer's recommendation.

Organic certification can be a valuable marketing tool for new, un-established farmers and poultry producers. It gives a certain feeling of safety to consumers and adds a readily recognizable branding to your products. Depending on the market and area, established producers may find that organic certification may

Feeding Pasture-Raised Poultry

not add value. Once you have established a trusting customer base, you may opt to continue organic practices but discontinue organic certification if your market demands or expectations don't consider it valuable.

Appropriate Feeder and Waterer Space and Position

Properly positioning and sufficient feeder and waterer space will save the lives of some of the poultry you raise. In general, the height of feeder and waterers should stay level with the average bird's back (the part of the chicken where the neck joins the main body of the chicken).

It is not only important to maintain the correct height of feeders and waterers, but also to not change the method used to deliver feed and water to the birds. For example, if you use red bell waterers in the brooder, or in the layer's winter housing, you should use the exact same bell waterers in the field. This avoids confusion for the birds, helps alleviate many potential stresses and minimizes growth and production losses. Changing feed and watering devices can cause losses of 2-3 days of optimal performance as the birds adapt. It can also cause death, if the birds don't adapt to the new systems.

Correcting feeder and waterer height may reduce feed waste and spillage on the ground by up to 25%. This can add up to a cost savings of up to $1 per broiler

Correct feeder height may reduce spillage by 25%

Feeding Considerations

and $2 per year per hen (based on conventional feed prices of $300 per ton. Cost of feed will double if you use organic feed.) Proper heights will also reduce water wasting and spilling on the bedding or ground, saving at least 10% on overall bedding costs. Controlling the waste and spillage of both feed and water will also minimize the potential flourishing of disease-causing organisms, and of problems associated with vermin.

It may seem natural for chickens to eat off the ground, as this occurs naturally. However, since processing grains exposes starches and carbohydrates, when they come into contact with moisture—spilled water, urine, and rainfall—they will begin to ferment. This fermentation takes place near feeders and waters, where there are also higher levels of manure. These conditions together create a perfect breeding ground for pathogens. Additionally, partially fermented grains have an almost addictive flavor to a chicken. Once they start eating partially fermented spilled feed, they will continue to look for these morsels on a routine basis.

These conditions lead us to a discussion of disease-causing organisms.

Disease-Causing Organisms

Clostridium botulinum is a Gram-positive, rod shaped bacterium that produces the neurotoxin *botulinum*, which causes the flaccid muscular paralysis seen in botulism. It is also the main paralytic agent in botox. It is an anaerobic spore-former, which produces oval, subterminal endospores and is commonly found in soil. Proliferation of *Clostridium* can occur on spilled grain, and the consumption of this grain by chickens can lead to botulism poisoning. By the time the symptoms are recognized, it is most likely too late for corrective treatment. Symptoms of botulism are lethargy, sleepiness, and limber neck, where the sitting bird's head will slowly drift forwards toward the ground or bedding. I refer to this as "Like falling asleep in church" or "Nodding off."

Clostridium perfringens is a Gram-positive, rod-shaped, anaerobic, spore-forming bacterium of the genus *Clostridium*. *C. perfringens* is very common in nature and can be found as a normal component of decaying vegetation, marine sediment, in the intestinal tract of humans and other vertebrates, insects, and soil. Virtually every soil sample ever examined, with the exception of the sands of the Sahara, has contained *C. perfringens*. *Clostridium perfringens* may or may not create symptoms similar to botulism. However, an excess will always cause necrotic enteritis, which is similar to an ulcer of the digestive tract.

Affected birds will go off feed and consume more water than normal. In most cases *C. perfringens* will cause a runny manure with some off coloring. If the symptoms are observed early enough, treatments may be effective. To treat a *C. perfringens* infection clean and remove the bedding or change pasture, and

Feeding Pasture-Raised Poultry

move feeder locations regularly. Offer whole milk or yogurt to the flock, approximately 1 pint per 100 birds, more or less depending on age or stage of development. Repeat this regimen for 7 to 10 days. Dairy products should be your first line of defense.

In more serious outbreaks, or those unresponsive to milk products, I have had good success using copper sulfate in the drinking water. Dose at 1 ounce copper sulfate per 5 gallons of water for 3 days. Use this as your sole water source. You may repeat this treatment as needed, after taking a break for 3 days without the copper sulfate. *C. perfringens* most often affects poultry between days 18 to 25, much like coccidiosis. The treatments are the same.

Coccidia are microscopic, spore-forming, single-celled parasites belonging to the apicomplexan class *Conoidasida*. Coccidian parasites infect the intestinal tracts of animals, and are the largest group of apicomplexan protozoa. Coccidiosis is a parasitic disease of the intestinal tract of animals. The disease spreads from one animal to another by contact with infected feces, or ingestion of infected tissue.[7]

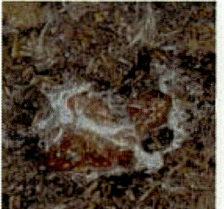

Manure of a coccidia infected bird

These organisms are not the only three poultry killers that are waiting for the high moisture, high carbohydrate (from feed), high manure environment in poultry litter or pens to proliferate (grow rapidly and over-populate). However, they are the three largest contributors to weakened immune systems and poultry death. *E. coli, Salmonella,* and *Camphylobactor* are waiting to join in the sick chicken party as well. The good news is if you maintain proper feeder and water height, and more than adequate access space you can improve the poultry's health, and improve growth performance which will reduce production time. [8]

Normal manure

Water Quality

Water is understandably necessary to sustain life. Poultry will consume twice as much water as feed by weight during normal growing conditions. During higher temperatures in summer months they may consume slightly more water, and during the winter months they may consume slightly less. Water quality and freshness are critical. If you wouldn't drink the water yourself, don't serve it to your birds. Don't just add new water to old–clean out water pans and start with

Feeding Considerations

fresh water daily for optimum results.

Water should be tested for mineral content. In some cases minerals found in the water source will influence the mineral requirements of the feed. Mineral content of water varies throughout the various regions of the world. It is common to find higher pH water in the upper plains states and Canada, while in the eastern U.S. there is commonly higher iron content.

The level of bacteria in water is also important. Excessive bacteria levels will cause poor weight gains, decreased rate of lay and may lead to higher rates of mortality. There are similar concerns for water nitrate or nitrite levels; levels in excess of 50 ppm will affect the performance of poultry. If either condition is found (high bacteria or high nitrates and nitrites), it can be corrected by chlorination, ultraviolet lights and filtration.[9]

8 years ago, now all grown up and no longer a flock manager

I often illustrate the importance of water quality by telling the story of my teenage daughter poultry manager. After many reminders about clean water and waterer cleaning, I continued to be severely frustrated with her management (as parents can be). I collected some water from the dirty, algae-growing waterer our backyard hens drank from, and proceeded to the house where I gave the ultimatum: "You will either drink this water from the chicken waterer or you will immediately clean and replace the water in the chicken's waterer." The response was priceless as she rushed out of the house to remedy the problem. That worked for a little while, but then new scare tactics were needed. I wish you luck with your teenage poultry manager!

Water Additives

Certain additives to water can help prevent or alleviate specific poultry issues.

• *Vinegar.* Apple cider vinegar, a mild acid, helps alleviate heat stress. It also has been shown to have a beneficial influence against heart attacks, ascites, and sudden death syndrome. Rate: one part vinegar to 100 parts water to start in an acute situation, drop back to one to 200 as maintenance. It is a fairly common practice in commercial poultry operations to add acid to the drinking water to help control many of the digestive tract illnesses such as coccidiosis and enteritis. There is also a belief that it has a positive effect on feed digestion efficiency.

• *Hydrogen Pyroxide.* Peroxide is a caustic substance and considered an organic alkali, which will raise the pH in the bird's digestive system. It is useful in the same way vinegar is. Rate: 50 ppm, measured at the furthest drinking device from the

Feeding Pasture-Raised Poultry

source. 50 ppm is approximately 1/3 of an ounce of store-purchased peroxide per gallon of water. If your water pH is low or acidic, use hydrogen pyroxide. If your water pH is neutral, vinegar is better.

• *Probiotics*. These are live beneficial microorganisms that can be fed to the birds. Probiotics will promote better gut health, increase feed efficiency, and will help to exclude harmful pathogens, such as *Salmonella, E-coli*, etc. Probiotics may be purchased as a supplement and fed according to manufacturer instructions when necessary.

• *Copper Sulfate*. This helps to treat coccidiosis and necrotic enteritis. Feed at the rate of one ounce per 5 gallons of water for 3 days only, as the sole water source.

• *Garlic*. Useful in fighting disease and overall infections. Rate: One clove of garlic cut lengthwise per 5 gallons of fresh drinking water. Allow water to sit overnight, to allow the allicin in the garlic to release into the water. In ancient medicine garlic was used as an antibiotic and antimicrobial treatment. It has been used for millennium in many forms, from fresh to tinctures.

• *Oregano Oil*. Oregano is fairly new to poultry producers, but shows signs of helping with ongoing brooder issues. Oregano naturally produces two antimicrobial compounds: carvicol and thymol. These compounds are commonly used in human products such as mouth-wash and toothpaste to help fight unwanted bacteria growth. Oregano oil, like many other essential oils, is very strong, and has caustic qualities. It should be used only as necessary and in small doses. It is best when using oregano oil or any other essential oil to get recommendations from a professional, as when used incorrectly they may have negative effects or even death.

• *Raw Milk*. Raw milk is a valuable all-round help in the brooder, where it helps to condition the chick's guts. The best time to use whole or raw milk is day four, after the chick has found the food and water sources. It is useful again on days 7 to 10 to help avoid coccidiosis and enteritis. Raw milk can be used as a treatment for many conditions, and fed free choice in a waterer for 5 to 7 days whenever you want to give the birds a boost. It also works very well as an appetite stimulant whenever birds appear to not be eating. Milk contains many simple sugars and fats that will restart a slowed digestive tract.

These management techniques are not the only factors that will determine how well your poultry thrive, but they are the most significant. You'll know when you get it "right" when you are growing Cornish cross broilers on pasture to 4 pounds processed weights at 6 ½ weeks of age (45 days) using less than 12 pounds of feed per bird. These production values can—and do—happen for outstanding pastured poultry producers on a repeated basis!

True Grit
The Lowdown on an Important Avian Supplement

Talking to hundreds of pastured poultry producers, I often hear:
"Why do my birds only grow to 3.5 pound carcass weight?"
"Why do my birds get necrotic enteritis?"
"Why is my feed conversion so high?"
These problems tend to have common denominators—one of which is the bird's access to grit.

I assumed everyone knew that birds needed grit. But when I started asking, "Are you using grit?" the responses I got amazed me:
"They get it from the ground."
"I feed oyster shells."
"They eat the gravel out of the driveway."
"Do they need grit?"

Few of the books available for small backyard-type producers talk about the need for grit, even though all of the large-scale producer manuals recognize the importance of offering different sizes of grit at different stages of development.

Nutrition of the Chicken, says on page 8: "Grit provides additional surface for grinding as acting to stimulate motility in the gizzard. The digestibility of coarse feed particles, such as whole grains, grain with minimal amount of processing, or pelleted feed, is improved by addition of grit in the diet."[10]

Morrison's Feeds and Feeding, agrees: "Grit aids in the grinding of whole grain or other coarse feed in the gizzard...Experiments have shown that grit should be supplied to laying hens, and that it is best to furnish hard grit in addition to limestone grit or oyster shell. In Ohio experiments, supplying layers with hard grit in addition to limestone grit or oyster shell increased the egg production 9.6 percent and reduced the feed requirement per dozen eggs 7.1 percent."[11]

Traditional Feeding of Farm Animals, states "...the partially digested feed is passed on to the gizzard, an extremely powerful grinding organ having a tough and convoluted lining, where it is ground to a very fine state by means of the abrasive action of the stones or grit which the fowl swallows. Pieces of glass have been taken from the gizzard of the chicken that were rounded on the edges and worn as smooth as though ground and polished by hand, and pieces of iron have been removed that had been bent double. Such instances give an idea of the

Feeding Pasture-Raised Poultry

toughness of the gizzard lining and the enormous muscular power of this natural grist mill."[12]

If grit is so important, what happens when it is missing? Missing or inappropriately sized grit can contribute to any of the three problems listed at the beginning of this section.

Why do my broilers only get to 3.5 pound carcass weight at 8 weeks?

Low carcass weights result when something keeps the birds from getting all the food and water they need, or keeps them from converting what they eat to muscle and bone. If your birds have adequate feeder and water room and enough high-energy, palatable feed, try adding grit. I have seen the addition of grit to a broiler diet contribute as much as ¾ to 1 pound of additional growth with no other significant changes to the feed supply or growing environment.

Why do my birds have a problem with necrotic enteritis?

Necrotic enteritis[13] is a bacterial imbalance in the gut. This can happen due to coccidial infection or injury to the intestine from the chick eating litter, or from spoiled feed containing harmful bacteria, or from feeds that are too finely ground (particularly those containing wheat). Feeds ground too fine tend to slow down feed passage through the digestive tract, causing an impaction. This impacted area in the digestive tract provides a low or no oxygen fermentation region, creating an optimum environment for pathogenic bacteria already living in the digestive tract. These bacteria take over the majority of the overall gut population, causing illness and gut wall lesions amd erosions. Grit helps keep the digestive tract moving forward, and reduces the possibility of impaction. Even the effects of enteritis can be reduced when grit is part of the diet.

Why is my feed conversion so high?

There are several potential reasons for poor feed conversions:
1. Improper feed nutrient levels. Keep in mind that a chicken always eats for its energy need, ONLY! It doesn't eat to meet its needs for protein, vitamins or minerals. When the bird's metabolism reaches the desired level of energy for the day's activities it will stop eating. Balanced feed formulas are a must for good feed conversions, for any species of animal.

2. Incorrect feeder and waterer height can lead to wasted feed. The edge lip of the feeder and water should be level/even with the average bird's back, the area just at the base of the neck. Broilers grow so fast that you'll want to be sure to check this several times a week and make adjustments as necessary.

3. Poor digestion, as mentioned above. Grit keeps the digestive tract moving.

True Grit

Grit also aids in proper grinding of the feed to the correct absorption size for better utilization. Grit also helps break up the more fibrous grasses and other plants, making them easier for the birds to digest–the whole reason for our putting chickens on pasture!

Grit is available in several sizes, appropriate for different species at different stages of growth: pigeon grit; quail grit; broiler starter, grower, finisher grit; layer grit and turkey starter, grower #1, grower #2, and finisher grit are all available. I am sure I missed a couple, but you get my meaning–there is specific grit for everything. All of the grit supplies that I am aware of are granite grits, one of the hardest stones known. The hardness determines the durability, how long it will last in the gizzard, and how effective it is in grinding food particles. Grit should be offered at day 2 or 3 of life and then continuously throughout the bird's lifecycle.

I have discovered that the more aggressive foragers require more grit. Turkeys are great foragers and will eat nearly 1/5th of their entire ration weight as grit. Turkeys will consume on average 60 pounds of feed during their grow-out. This means they'll use 12 pounds of grit or grinding material. I often joke with folks that a turkey will eat its weight in grit if it has free access. Even a modern breed broiler can consume 1/20th of its diet as grit. Depend on the species' foraging instinct to help determine how much grit your poultry will eat. Have grit available at all times; your poultry will eat what they need. Why else would they eat stones?

Limestone and oyster shell are excellent sources of calcium, but they are pretty soft. Chickens using them for grit will need a lot more of these items than they will granite grit. Limestone and shells are also more expensive than grit, and will take up valuable room in the bird's gut needed for feed.

I often hear, "They get all the grit they need from the pasture/dirt." Try this test: dig up 4 square feet, 1 inch deep of the surface soil where your chickens have access. Now screen this topsoil for the stone particles in the soil. Sort and grade the particles into piles of the size appropriate for your poultry's stage of growth. Unless your chickens are grazing a gravel pit (in which case there wouldn't be much grass growing), you're unlikely to have enough grit in the pasture for your birds. If your local feed supply doesn't offer the size you need, insist on their ordering it. Grit is cheap and it never goes bad, so even buying a year's supply will not cost much.

Consider offering whole wheat in a separate feeder mixed with grit, in a ratio of 4 parts wheat to 1 part appropriately sized grit. This will encourage birds to eat more, increase feed availability and cut down on prepared feed expense. This should not cause decreases in weight gains or feed efficiency, and the birds love it! I start this with turkeys at 8 weeks, broilers at 5 weeks and layers throughout the summer when on good range. They will all balance their own diets pretty well when given the opportunity.

Feeding Pasture-Raised Poultry

Feed Sources and Preparation

Commercial Feeds

Commercially produced feeds are available from all of the major manufacturers (i.e. Purina, Master Mix, Agway, and others). Commercially produced feeds will contain ingredients such as corn, soybean meal, meat and bone meal, feather meal, porcine meal, wheat mids, bakery meal (a byproduct of the bakery industry), vitamins, minerals, and preservatives.

When using commercially available feeds it is difficult for the processing, transportation, purchasing and then consumption by the animals to occur within a 30-day timeframe. Although variable with the feed and time of year, commercially processed bagged feeds will be on average at least 6 months old by the time they are consumed. If feeds are manufactured at one central location, warehoused, and then distributed throughout a large area to smaller retail outlets they may need a 6-month shelf life.

At the retail outlet they may be stored an additional 30 to 60 days prior to sale to the consumer. For this reason, preservatives will often be added to feed to minimize oxidation of the fats and carbohydrates. Flavor additives provide an appealing aroma and flavor to old feed. These additives do not, however, preserve the freshness or digestibility of the feed from the bird's perspective.

Commercially produced feeds are always produced under least-cost guidelines. Commodity values change every day on the Chicago Board of Trade market. Professional grain brokers are paid large salaries based on their ability to buy commodities at the most advantageous times. They spend their entire day watching a computer screen for significant price movement on potential feed ingredients.

A rise in the price of a specific protein could cost a large feed producer thousands of dollars. So, feed ingredients are regularly changed in response to commodity price changes. This is "least-cost rationing"–and means that the ingredients of tomorrow's ration may be significantly different than those in today's ration.

Since the feed is delivered in unrecognizable forms (pellets and crumbles), the feed will still look the same each time you purchase it–even though ingredients may differ significantly. Least-cost feed rations create problems for producers who are trying to effectively manage their flocks, as the exact nutritional value will change from batch to batch of feed purchased.

Feed Sources and Preparation

Crumbles vs. Mash Feed

The debate of pellets or crumbles vs. mash feed has gone on forever. Either crumbles or mash are suitable for poultry. Pellets and crumbles were designed to provide uniformity of nutrients in every morsel of food. Some studies lead us to believe that the heating and steam process of pelleting in some way makes the feed more digestible.

Crumbles start as a pellet that is then reprocessed through a cracker or roller mill. Since small mills that do custom rations for small growers usually do not have pellet or crumbling capability, their rations are usually in mash form. Many people feel that a mash diet is better because growers can see the ingredients and have control over them. My mentor, Jack Robinette, started me out recommending mash-style feeds. He believed that if the mash feed was processed correctly, and the particle size was right, that pelleting and crumbling simply added unnecessary expense. (Pelleting will add $30 to $50 per ton of feed.) He also felt that pelleting can be used to hide by-products and less palatable ingredients.

Mash particle size should be large enough to be able to easily identify 75-80% of all of the grains used. The photo to the right shows acceptable mash particle size.

Bagged vs. Bulk Feed

The differences between bagged feed and bulk feed are:

• *Price:* Bagged feed is more expensive than bulk due to added cost of production and packaging. If this is not true at your mill you should be wary of the ingredients of the bagged feed. The ingredients may not be of the highest quality.

• *Convenience:* For growers with groups of birds over 500 at one time, bulk delivered feed will be much easier if it's available. You can get a custom ration made for you. Growers with under 500 birds generally opt for bagged feeds, usually using the feed mill's standard formulation. Bagged feed allows the grower to buy

Feeding Pasture-Raised Poultry

small quantities at any time, and as long as there is enough turnover of stock at the mill, helps guarantee freshness.

• *Freshness:* Freshness of feed impacts the appetite—and therefore the productivity—of poultry. When the feed is fresh, poultry will eat more overall, and will eat more frequently. Feeds are at optimum nutrient levels for up to 14 days after grinding, and maintain satisfactory nutrient levels up to 45 days after grinding or milling. After 45 days the feed is generally so stale or oxidized that poultry appetites will be severely depressed.

Oxidation starts immediately after the grinding or cracking of the grain. This oxidation occurs as a reaction between newly exposed moisture and starches within the grains. The oxidation of the starches will cause some energy loss of the feed value. However, in most cases these losses are negligible if the feed is fed within 30 to 45 days of processing. Typical nutritional losses under normal storage conditions for 60 days include: 10% of energy, 1 to 2% of the protein, 10% of most vitamins, 30% of vitamin K, and 15% of riboflavin.

Milling

Several milling options are available for poultry producers.

• *Large-scale feed mills:* These mills only deal with larger customers capable of receiving bulk feed in larger quantities. They generally require minimums of 8 to 10 ton deliveries. Typically larger feed mills deal mostly or exclusively with contract growers or corporations with very large poultry populations.

• *Small-scale feed mills:* These mills generally cater to the small farm units that wish to receive 1 to 2 tons per delivery. These mills have minimum mixing requirements for special orders ranging anywhere from 500 to 2,000 pounds. The minimum mixing requirements may be partly due to machinery size, or to the fact that the mill chooses to not be bothered with orders less than 1 ton.

Feed Sources and Preparation

• *Farm-scale milling:* This is the optimum for feed quality, as you can make feed as needed to maintain freshness. Farm milling is generally done with grinder/mixer combinations. There are several manufacturers listed in the reference issue of *Feedstuffs* magazine (see References). I find used grinder mixers of various sizes and brands at local farm auctions in my area will sell from $400 to $4,000. These values, of course, depend on the mill's model, capacity and condition.

New grist mills, both manual and powered models, are available from:

The C.S. Bell Co.
170 West Davis Street, P.O. Box 291
Tiffin, OH 44885
419-448-0791 www.csbellco.com

Mills are also available through local dealers representing their products.

Soybean Roasting

Soybeans must be roasted or heat-treated other ways for poultry feed in order to breakdown a trypsin-inhibitor enzyme that interferes with protein digestibility and thus influences the bird's growth. A urease test should be done after roasting to determine if the enzyme breakdown is sufficient. There are many different opinions on the methods and temperatures required for the roasting process. The most common recommendation for roasting whole soybeans is heating to 270 degrees F and holding them at that temperature for a minimum of 20 minutes.

The list of companies that offer small-scale roasters is short:

Gem Grain Roasters – Propane or natural gas fired, fixed location
W21596 Erickson Lane
Arcadia, WI 54612-7211

Schnupp's Grain Roasting, Inc. – Propane or natural gas fired, portable
R.D. 6, Box 840
Lebanon, PA 17042
800-452-4004, fax 717-865-7334
Sells the Roast-A-Matic

Dilts-Wetzel Manufacturing Co. – Hot oil, electic heated, stationary
2501 W. Washington Rd.
Ithaca, MI 48847
989-875-4069, fax 989-875-4589
sales@diltswetzel.com

Feeding Pasture-Raised Poultry

Another common method of heating soybeans is called extruding. In this process the beans are ground to a meal consistency–and then forced through a small die or orifice. The friction of forcing the meal through the orifice generates the heat required to inactivate the enzyme.

Feed Storage

If you know you'll have to store feed for a prolonged period, you can slow the rate of oxidation by improving the storage method and conditions. When ground feed is stored in limited air (or airtight) containers, minimal oxidation can occur. Generally unprocessed grains will store very well and hold their nutrition for up to two or three years. Storage requires water proof, bug proof containers or bins, preferably in cooler conditions. For optimum storage, the use of airtight containers is best but not necessary. Storage of most grains for extended periods of time may require treatment with an insecticide to repel insects. Grains should be dried to 12-15% moisture content prior to storage to avoid condensation and molding.

Feed Quality Issues

Molds and Toxins in Feed

In recent years there have been warnings sent out about high vomitoxin contents of wheat and barley, and also corn and corn silage. If corn doesn't dry down correctly, and the ears stay upright, rain can get between the husks and kernels. The kernels are not able to dry out correctly, allowing mold formation in the ears.

During the spring, summer and early fall, I rarely think about molds and toxins because the ready availability of pasture and fresh forages seem to negate the effects of stored feed molds and toxins. However, during the winter months, when we feed all stored or fermented feeds to our birds and other animals, the effects of molds and toxins become much more prevalent. Molds and toxins can have very negative effects on a bird's performance and health, and thus your profitability. We'll review those effects later in this section.

Often when I ask farmers if they would like to do a mold and toxin test, the response is, "No, it looks fine," or "I don't see any, so it should be ok." Visual examination or smelling of the grain often leads us to believe there are no molds or toxins present. Surprisingly, molds and toxins are not always visible to the human eye or nose. I have seen samples submitted to the lab that look and smell great, but when the test comes back, the samples have tested positive for mold growth and toxin formation.

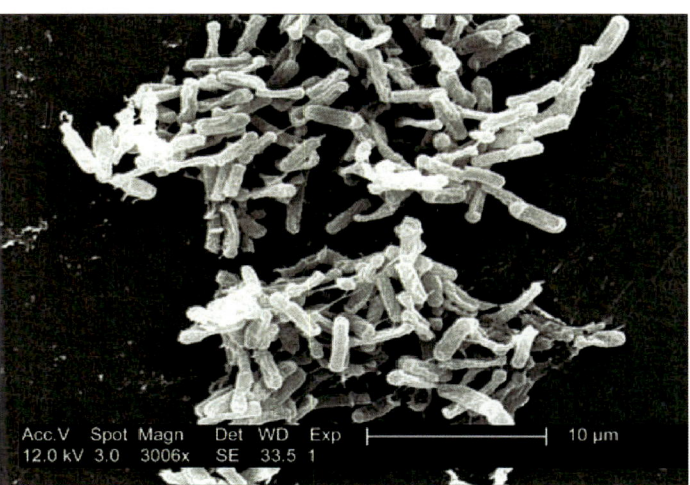

Clostridium difficile under 3006x magnification Using a 10 micrometer scale. Not the human eye!

Feeding Pasture-Raised Poultry

I am not suggesting that you need to test all grains every year. However, if you see one or more of the following symptoms in your animals, you should get your grains and forages tested:

- Decreased appetite or feed refusal
- Poor weight gain
- Decreased reproductive or laying performance
- Suppressed immune system
- Respiratory infection
- Almost any unusual symptom that you do not normally see

A mold and toxin test will cost about $80. This may seem expensive, and none of us likes to spend money unnecessarily. But, if a test exposes moldy feed, and then saves you ten market weight broilers, or is the difference of half the saleable carcass weight of your flock, it is money well spent.

Another way to approach the problem is to feed corrective supplements in anticipation of any mold and toxins. Bentonite is the cheapest mold and toxin binder on the market. However, if you feed bentonite or other clay products when molds and toxins are not present, the clay particles will bind to beneficial nutrients, rendering them useless. Both the bentonite and the nutrients it has bound now become very expensive fertilizer.

You will experience production loss at mycotoxin levels of 2 ppm (parts per million). Two ppm is the equivalent of 2 grains of sand in a half-gallon jar of sand.

(A grain of sand weighs on average .0026 grams. There are 453.6 grams in a pound. That means approximately 174,461.5 grains of sand in a pound, and 5.732 lbs of sand for 1,000,000 grains of sand.)

So, the equivalent of two bad grains of sand in 5.732 pounds of sand are enough to decrease feed efficiency by 10%.

If a finished broiler consumes on average 16 pounds of good feed, using "slightly" contaminated feed means it will take 17.6 pounds to finish one bird. If feed costs an average of 25¢ per pound, then your costs rise by 40¢ per bird. If you raise 1,000 broilers and have 2 ppm mycotoxin contamination—you lose $400 in increased feed costs alone.

I recommend that you add up all your potential losses connected to mold and toxins. Then do yourself, your animals, and your checkbook a favor and check your feed for molds and toxins!

Evaluating your Feed Cost

How you raise your chickens is a personal conviction based on your beliefs. I am not here to say any one way is better than the other. You, the producer, must decide where your priorities are and what you are willing to consume and market to others.

However, I often hear comments such as "That's too expensive," or "I can't afford to feed organic feed," or, best of all, "It doesn't make a difference what I feed."

These kinds of comments and conversations stimulated me to calculate exactly how much organic versus non organic feed costs. In the next few pages is the answer to "Just how much more does it cost to feed organic feed?"

In 2009 I gathered prices from six different feed suppliers in different regions of the country. Three of them are conventional local feed mills and three of them are organic feed mills. To have a fair comparison, I chose suppliers who all used our Fertrell Poultry Nutri-Balancer. Although prices have changed since 2009, the relative values have not changed that much, and so the comparisons (and methods used to do the calculations) are still valid.

Figuring the Cost of a Dozen Eggs (2009 numbers)

Price per pound of feed x the amount of feed consumed per hen = the cost per hen per day. Multiply cost/hen/day by 100 for the cost per 100 birds.

Divide the cost to feed 100 birds by a factor of 70–which is the lay cycle average production for the flock. This yields the cost per egg at 70% production, which is what you can expect in an average operation.

Take the price per egg, multiplied by 12 for a dozen eggs, (or by whatever number your sales container unit is) for your final cost of feed per dozen eggs (or other sales container unit).

Example: feed cost 24¢ lb. x .28 lb. feed/hen/day = 6.72¢ cost/hen/day.
 6.72¢ x 100 hens = $6.72 cost per 100 birds
 $6.72 ÷ 70 production level = 9.6¢ cost of a produced egg
 9.6¢ x 12 = $1.152 per dozen feed cost.

Feeding Pasture-Raised Poultry

The chart below shows the calculated cost based on the formula provided above for feed from several feed suppliers in 2009. (Note that any transportation costs are not included in these calculations. Obviously you will either have to pick up feed or have it delivered–either way there is a cost you will have to include. Many farms find that they have to go further afield to find organic feed, and so the organic transport may add more to your cost. The cost per ton or mile, however, should be the same whether you are hauling organic or non-organic feed.)

I have often heard that organic production costs about twice as much as conventional production. These calculations prove those claims to be true.

Feed Manufacturers: Fertrell Formula Ration, Layer Feed
(2009 numbers)

	Layer Feed ¢/Lb	Feed Cost $/Dozen
Organic Feed-		
Organic Unlimited, Atglen, PA	$ 0.48	$ 2.31
Lakeview Organic, Penn Yan, NY	$ 0.40	$ 1.92
S & S Grain, Acadia, WI	$ 0.34	$ 1.61
Conventional Feed-		
Ross' Feed & Grain - GMO Free -Quarryville, PA	$ 0.24	$ 1.15
C.E. Sauder's, East Earl, PA	$ 0.19	$ 0.91
Fertrell Formula, Holmes Ag - Holmesville, OH	$ 0.18	$ 0.84

Figuring the Cost of Raising a Broiler

I did a cost evaluation for broiler feed cost using a similar formula. The same feed suppliers provided 2009 prices for Broiler Starter-Grower feed so that I could calculate the average feed cost per broiler.

The formula is simpler for broilers:
Carcass weight of the broiler x three pounds of feed per pound of carcass = total pounds of feed consumed per bird. Multiply the total weight of feed consumed per bird by the cost per pound of feed to get the cost of feed/bird.

Example: 5 lb carcass weight broiler x 3 lbs of feed per pound = 15 lbs of feed consumed to get to market weight.
    15 lbs x 25¢ lb. feed = $3.75 feed cost for the broiler.

Evaluating Your Feed Cost

Feed Manufacturers: Fertrell Formula Ration, Broiler Feed
(2009 numbers)

	Broiler Feed ¢/Lbs	Feed Cost $ /Broiler
Organic Feed-		
Organic Unlimited, Atglen, PA	$ 0.52	$ 7.80
Lakeview Organic, Penn Yan, NY	$ 0.42	$ 6.30
S & S Grain, Acadia, WI	$ 0.35	$ 5.25
Conventional Feed-		
Ross' Feed & Grain - GMO Free, Quarryville, PA	$ 0.2495	$ 3.74
C.E. Sauder's, East Earl, PA	$ 0.2208	$ 3.31
Holmes Ag - Holmesville, OH	$ 0.1800	$ 2.70

Most of the other costs of production (labor, marketing, etc.) will be the same whether you raise organic or conventional poultry. However, processing costs may be higher if a certified organic processor is used, and transportation to a certified processor may be farther and thus more expensive than to a relatively more commonly found non-organic processor.

This information should help you evaluate what your costs of production are—or could be. After you have made your production decisions based on your personal convictions and beliefs, and you have explored what your market will pay for a higher quality product, you can use the equations to assess your cost of production. All of these things together should guide you in setting your final market price.

Feeding Pasture-Raised Poultry

Poultry Feed Ingredients

Common Ingredients Used in Poultry Feed

Corn

Corn is primarily used to supply energy to the diet. Other benefits of corn are the yellow/orange pigmentation, with xanthophylls (5 ppm) and carotenes (0.5 ppm) enhancing yellow skin and fat coloration. Corn has no feed inclusion limitations. Corn should be "medium ground" to uniform particle size, slightly smaller for chicks and larger for adult poultry. Corn is a staple ingredient in poultry diets. Corn grown in southern regions has a higher potential of aflatoxin contamination, as well as many other toxins, than other cereal grains. The formation of aflatoxin molds which lead to toxins are generally brought on by plant stress during the growing season. To avoid aflatoxin, try to buy high quality grains from areas without a difficult growing season.

Milo, Grain Sorghum

Milo is a very effective substitute for corn. It has between 7-10% protein and energy between 1300-1500 Kcal per pound. Up to 50% of the corn in a diet may be replaced safely with milo. Milo contains a chemical compound known as tannin. High tannin milo varieties were developed to discourage wild birds and animals from eating the grain while it is growing. High tannin levels will also inhibit poultry feed intake. Tannins are easily recognizable by the red or rusty color of the grain. Lighter or white varieties of milo are better suited for feeding poultry, as they are lower in tannins. A 1% rise in tannin content reduces overall grain digestibility by 6 to 7%.

Red Milo

White Milo

Poultry Feed Ingredients

Wheat

Wheat is commonly used in many countries as a major source of feed energy. Wheat is higher in protein than corn or other small grains (9 to 16% protein, 1440 Kcal/lb energy). Hard, red spring wheat varieties tend to have higher protein than soft, white winter varieties. Although birds love it, wheat should be limited to 30% of total feed by weight unless enzymes are added to aid digestion. For proper digestion of wheat you should add xylanase enzyme, following manufacturer's directions. Lysine should also be added, as wheat is low in this amino acid.

White Wheat

Turkeys may be offered wheat free choice mixed with grit, 4 parts wheat to 1 part grit from week 9 until they are finished growing. This should be offered in a separate feeder in addition to the processed grain mix.

Red Wheat

Broilers and roaster meat birds may also be offered wheat in an additional feeder at the same rate of 4 wheat to 1 grit after week 5 of development.

Triticale

Triticale is a cross between wheat and rye. It is a very high-yielding small grain and easily accessible in many cooler climate regions. Triticale is generally 11 to 14% protein with energy between 1300 to 1400 Kcal per pound. Triticale should be limited to between 15 to 20% of the diet, depending on the age and type of poultry being grown. This limitation is due to high pentosans content. I am not aware of an enzyme that can be added to offset the effects of the pentosans. These same restrictions are also true for feeding rye cereal grain.

Oats, Barley

Oats and barley add fiber and increase the bulk density of feed. The hull makes up 20% of the weight of oats and barley. High fiber from these small grains keeps the digestive track clean, and can also be used to limit feed intake. High-fiber small grains also add protein and energy, although most of this added nutrition

Feeding Pasture-Raised Poultry

is burned off digesting the excess fiber. Oats and barley have an inclusion limitation of 15% by weight in any combination without added digestive enzymes (β-Glucanase). The result of excess oats or barley is wet litter and poor digestive viscosity.

Peas: Field, Cull, Black Eye, Cow, etc.

Peas are a commonly used protein source, particularly in cooler regions where soybeans are not easily grown. Peas have protein content between 22 to 24% with energy between 1100 to 1300 Kcal per pound. Peas should be limited to 15% in younger poultry, but may be fed as high as 25% of the diet in mature poultry. These limitations are due to high tannins in some varieties as well as low sulfur amino acid content (methionine and cystine). Methionine is poultry's first limiting amino acid, responsible for proper feather development and essential for proper immune system development. Lower tannin level varieties of peas should be used for better digestibility.

Oats

Barley

Austrian Winter Peas

White Peas

Green Peas

Poultry Feed Ingredients

Fishmeal

Fishmeal provides a varied form of concentrated protein. Fishmeal also helps balance all of the essential amino acids, most importantly methionine and lysine. Fishmeal will stimulate the bird's appetite, as poultry have an instinctive craving for meat proteins. Fishmeal averages 58 to 72% protein and 1280 to 1550 Kcal/lb of energy. It is generally made from herring, menhaden and anchovies (high-oil varieties) or from farm raised catfish (low-oil). Beware of catfish meal, as it has a potential for higher levels of contaminants, antibiotics or heavy metals.

Menhaden, commonly known as "Bunker"

Crustacean Meal

Crustacean meal is generally produced through the processing of either crabs or lobsters. After the large portions of meat have been removed from the crustacean, the residual is then ground and dried for animal feed and fertilizer use. Crustacean meal can range anywhere from 25% to 45% protein, with a low energy of 800-1000 Kcal per pound. While very beneficial in delivering a meat-type protein molecule (that is, the protein is higher in methionine than a grain protein), the volume of crustacean meal that can be incorporated into the ration is limited by the amount of naturally occurring salt it contains. It is also limited for use in a poultry diet due to the lower energy value. Interestingly, the composition of sea crus-

Blue Crab

Crab Meal

Feeding Pasture-Raised Poultry

tacean is almost identical to the nutritional composition of dry land crustaceans (for example crickets and grasshoppers).

Vitamin/Mineral Premix

The primary purpose of a vitamin/mineral premix is to balance vitamins, macro and micro minerals to meet modern poultry health and performance needs. There are various formulations available, of differing content quality and value.

In 1996 the Fertrell Company developed Poultry Nutri-Balancer, a vitamin and mineral premix specifically formulated for poultry. Poultry Nutri-Balancer was originally formulated to meet the needs of organic producers, where antibiotics and chemicals can not be used to cure health problems.

Working with my mentor Jack Robinette, we decided to put extra vitamins, trace minerals and kelp meal in the formula as an insurance plan. Most of the vitamin levels are at least 10% higher than the National Research Council (NRC) recommended requirements. For some vitamins that are easily oxidized, such as B vitamins, we set levels as high as 500% of the NRC recommendations. Poultry Nutri-Balancer has proven to be extremely beneficial for both organic and conventionally fed flocks and all species and classes of poultry. (For more on the benefits of Poultry Nutri-Balancer, see Appendix A).

Poultry Nutri-Balancer is now essentially the same formula that we developed in 1996. The only exception has been the addition of direct fed microbials and probiotics–which where added after many requests from producers.

Poultry Nutri-Balancer is currently used in 40 states, Canada, Panama and Bermuda. We have experienced tremendous success with it over the past 17 years. There is enough Poultry Nutri-Balancer used across the United States and Canada to feed more than 2,000,000 broilers, 330,000 laying hens, or 500,000 turkeys. That's a lot of pastured poultry!

Amino Acids

Methionine and lysine may be added to rations to balance amino acids, the building blocks of protein. Amino acids display a prefix L- or D- that describes the structure of the molecule (isomer). Amino acids occurring in animal tissues are always L- isomers. D- isomers have no biological function in animal tissue. Methionine is an exception to this rule. Poultry are able to use both D-methionine and L-methionine.

Depending upon geographical location and grain availability, additions of methionine or lysine may be required.[14] When poultry are fed an all vegetarian diet methionine **must** be added to meet the minimum nutritional requirements for

Poultry Feed Ingredients

proper development and growth. Lysine is typically needed in diets that do not contain corn and/or soy.

Salt

Salt is required to support normal body functions and electrolyte balance. The salt requirement for poultry is not very high. Generally speaking, 6 to 7 pounds of salt in a 2000 pound batch of feed is sufficient. In some regions where the ground water has higher levels of naturally occurring sodium it may be necessary to lower levels of salt added to feed. Test your water source to verify the mineral content and quality.

Probiotics or Direct Fed Microbials (DFMs)

Beneficial bacteria added to the diet aid digestion and nutrient absorption resulting in faster growth and better overall health. Direct fed microbials also replenish the beneficial bacteria flora, those "good bacteria" which directly compete with destructive or harmful pathogens such as *Coccidia, E. coli* and *Salmonella spp.* Excess beneficial bacteria are excreted in the manure and this helps to correct the balance of bacteria in the litter.

Soybeans

Soybean Meal (solvent extracted): Soybean meal is the standard protein used by the poultry industry. The meal is a by-product of the vegetable and industrial oil industry, which removes the valuable oil, leaving the high-protein meal which is used for livestock and poultry feed. The amino acid profile of soybean meal is well suited to poultry nutritional needs when combined with corn or sorghum.

The original bean has 18% oil. After oil removal, only 1.5% oil remains, hence decreasing the energy value from that of the whole bean. The oil extraction method includes de-hulling, cracking, conditioning to 158 degrees F, flaking to 0.25 mm, and then adding hexane to enhance oil extraction. Hexane must be removed from the meal because it is highly combustible and a potent carcinogen. There is no limitation on hexane for feed inclusion, except when extremely high levels of soy meal are used in rations such as turkey pre-starter. Too much residual hexane will lead to reduced carbohydrate digestibility, with side effects that may include wet litter and foot pad lesions in poults.

Raw Soybeans: Raw soybeans should not be fed to poultry. When used for feeding poultry, soybeans must be heat-treated to destroy a trypsin inhibitor enzyme (urease) that blocks protein digestion. The presence of the trypsin inhibitors will also cause an enlarged pancreas (a 50% to 100% increase over normal). These side effects will occur at raw soy inclusion levels as low as 5% by weight.

Feeding Pasture-Raised Poultry

Roasted or Extruded Soybeans: Roasted or extruded soybeans are an excellent source of both energy and protein. Typically the protein value for a roasted or extruded soybean is 36 to 40%, with energy of 1450 to 1600 Kcal/lb, depending on the oil content of the seed. This energy is a non-carbohydrate type energy that does not contribute to gout in poultry.

Roasted soybeans must be heated to 270 to 300 degrees F and held at that temperature for 10 minutes to ensure breakdown of the trypsin inhibitor. Whole roasted soybeans should be "medium ground" to maintain uniform particle size in relation to the corn and other ingredients in the ration.

Extruding means grinding, heating to 270 degrees F and then using mechanical pressure extraction to remove the oil. This process uses a screw auger within a mesh casing that traps the meal and separates the oil. The entire operation is encased within a steam jacket to allow the oil to flow more freely, and to provide the heat required to break down the trypsin inhibitor.

Extruded soybeans will be in a meal form when purchased, and no further grinding is necessary. Extruded soybeans should not be stored for more than 30 to 60 days (depending on time of year, heat and humidity) prior to use. The oil of the soybean has been exposed during processing and will oxidize and turn rancid over time. Rancid soy oil smells like old motor oil.

The extruded soybean meal is suitable for poultry and livestock feeds without feeding limitations other than excess protein. The extrusion process leaves approximately 7% of the original oil in the meal. The extruded meal provides an energy value midway between roasted soybeans and solvent-extracted soybean meal. Extruded soybean meal is a very suitable protein source for poultry raised in warmer climate, because these birds have lower energy needs than birds raised in colder climates. However, extruded soybean meal is not available in all areas. It is most commonly available in areas where there is high dairy production.

Neither the roasted nor extruded forms of soybean have feeding limitations.

Roasted Soybeans vs. Soybean Meal: Roasted soybeans' natural oil is easy to digest and will also generate heat during digestion, which warms the bird. Roasted soybeans have a good smell and flavor, and poultry eat them well.

On the other hand, solvent-extracted soybean meal has solvent (hexane) residues. Vegetable oils must be added to the solvent extracted meal to correct for the energy lost from the extracted oils. These vegetable oils added after extraction do not create similar heating when digested.

Poultry Feed Ingredients

Less Common Ingredients Used in Poultry Feed

Sprouts

Sprouted cereal grains may be used to increase vitamin content, especially carotenes. There are numerous studies regarding the feeding value of sprouted grains, and results of these studies are contradictory. Sprouts may also be used as a source of year-round forage.

"For a time, sprouted oats were used to a considerable extent in the winter feeding of poultry to furnish a green and succulent feed. With the recent advances in knowledge concerning the importance of vitamins and other factors in poultry nutrition, efficient rations have been developed that have made the labor and expense of sprouting oats unnecessary. Therefore the practice has been largely discontinued."[15]

The feed value of grains changes significantly when sprouting occurs. Studies show decreased energy value, increased protein percentage and increased vitamin percentage. The reason for such changes in nutritional values of sprouted grains occurs because as life begins, the energy from the seed is utilized very quickly to sustain the life of the new sprout.

Wheat Sprout Analysis Study
Monogastric Ruminants

	Moisture	Protein (As Fed)	Protein (Dry Matter)	Energy Kcal/Lb
Wheat	12.40%	11.10%	12.70%	1382
Day 1	46.50%	6.70%	12.50%	835
Day 2	60.10%	5.20%	13.20%	623
Day 3	65.00%	4.80%	13.70%	541
Day 4	72.60%	3.80%	14.00%	420
Day 5	77.90%	3.30%	14.80%	332
Day 6	83.90%	2.80%	17.20%	237

Research done by Gregory McCallum, 32 Geist Road, Lancaster, PA 17601

Dairy Products or By-Products

Many studies show the benefits of feeding milk and milk by-products to poultry. Most studies recommend feeding rates of between 2-5% of the diet.

"Skim milk and buttermilk for poultry: Dairy by-products are especially valuable for poultry, and most commercial poultry men use rations including some milk by-product. Not only does milk furnish excellent protein, but also its high content of riboflavin is of particular value for poultry."[16]

"Whey has only about one-third as much calcium and phosphorus as skim milk... It is nearly as rich in riboflavin as is skim milk. Whey is much more watery in composition than skim milk. When feeding whey it is very necessary to bear in mind

Feeding Pasture-Raised Poultry

the fact that most of the protein has been removed, and that the whey is not a protein-rich feed, like skim milk or buttermilk."[17]

"Value of milk in poultry rations: A comprehensive three-year study at Pennsylvania State College in which additions of dried skim milk were added to a high grade all mash ration showed that, when adjustments were made to keep the protein, calcium and phosphorus contents uniform, the rate of growth during the first two to four weeks of age, total feed intake, gain of weight during the growing period, and feed efficiency in the early part of the growing period all increased with increasing amounts of dry skim milk in the all mash rations. Most efficient and economical gains in up to four weeks were made with 2.5 and 5 percent milk. Greatest extra gains in weight per pound of dry skim milk consumed from 4 to 12 weeks occurred in the groups fed 1.25 to 2.5 percent dry skim milk… The University of Maryland reports that dried skim milk exerts a greater stimulating effect on a chick growth than can be explained on the basis of its riboflavin content. This is not true of dried whey."[18]

Morrison recommends feeding skim milk and buttermilk to laying hens. There are no limitations on the amount that can be fed, but generally 12 to 14 quarts per 100 hens is sufficient.

I personally reserve the option of feeding milk products for treatment of coccidiosis and necrotic enteritis. The soothing affect of milk fed during these conditions enables the affected poultry to resume eating and drinking normally. Feeding milk during these times also enhances the growth and reproduction of beneficial bacteria in the digestive tract, providing competition for harmful bacteria and organisms.

I believe that raw or clabbered milk has an even greater digestibility and will promote better gut health than the processed milk products used in these studies. Limited amounts of milk or milk products can definitely enhance poultry performance. However, unlimited or excessive amounts of milk products can be counterproductive. I have seen broilers and other species consuming free-choice or force fed high levels of milk with enlarged, discolored and fatty livers.

Pasture

As this publication is written primarily for the benefit of pastured poultry producers, it is necessary to understand the importance of pasture to the diet. It is evident that poultry do consume green growing plants. Each type of poultry will consume different levels of forage.

"Extensive experiments have been conducted to compare good pasture or range, with confinement or bare range for growing pullets or for laying hens. These experiments have shown that pullets which have been raised on fresh, uncontami-

Poultry Feed Ingredients

nated pasture are usually more thrifty than those raised in confinement...Numerous experiments have shown clearly that the farm flocks of small or moderate size, it is more economical to provide clean, uncontaminated pasture during the growing season than to keep the layers in confinement."[19]

Modern broiler breeds have very little desire to consume plant vegetation. However when provided with high quality forage, we observe as much as 20% of the diet intake from forage. We see this mostly on forage such as clover and alfalfa, maintained at an immature growth stage, from 6 to 8 inches tall. It appears that poultry prefer legumes over other types of plants. Gathering data from year to year and producer to producer, I conclude that pastured poultry eat 5 to 20% of their total diet as pasture, depending on type and age of poultry and quality of forage growth.

Feed efficiency depends on feed concentrate intake, water intake, live weight, and average ambient temperatures. A typical industry broiler in a temperature-controlled environment will have a feed conversion rate of approximately 2.09 pounds of feed to 1 pound live weight. Climate differences and temperature fluctuations make it difficult to accurately estimate feed efficiency on pasture.

Insects

Personal observations have led me to the conclusion that insects are of great benefit in poultry diets. They are extremely high in protein, with proteins and amino acid profiles comparable to those of fish meal or meat meal. Unfortunately there is little data on the nutrient content of insects. Below are some nutrient values from the websites listed:

Insect	Protein %	Fat%	Sources:
Crickets	6.7	5.5	www.planetscott.com/babes/nutrition.htm
Termites	14.2	NA	
Caterpillars	28.2	NA	www.eatbug.com/
Weevil	6.7	NA	http://ohioline.osu.edu/hyg-fact/2000/2160.html
Large Grasshopper	14.3	3.3	
Silk Worm Pupae	9.6	5.6	www.riverdeep.net/current/2002/03/030402_eating-bugs.jhtml
Giant Water Bugs	19.8	8.3	
Very Large Spider	63	10	

"...a certain minimum amount of feeds of animal origin should be included in the ration. In the case of poultry that are confined, there is a greater benefit from including in the ration such supplements as meat scrap, fish meal, or dairy by-products, than there is for poultry which are on good pasture. This is due both the quality of protein in good pasture forage and to the worms and insects they secure on pasture."[20]

Vegetarian Diet vs. Meat Products Diet

Both meat-based and vegetarian poultry diets can be formulated to meet proper requirements. Concerns have arisen regarding consumer health and disease potential from the use of meat and bone meal or other animal by-products. These are valid concerns, and I recommend the avoidance of re-feeding of domestic animals.

The pitfall to vegetarian poultry diets is that in nature poultry eat non-plant foodstuffs. These foodstuffs include insects, small reptiles, and even small rodents. There is definitely a place in a poultry diet for meat-type protein whether we provide it or they catch it. Meat-type proteins will more easily meet the bird's amino acid requirements.

Vegetarian diets (not containing any meat-type proteins) will require additional balancing of methionine, which is not a prevalent amino acid in plant protein. I personally feel that diets that include fishmeal (in place of domestic meat meal) reduce the possibility of cannibalization and will better satisfy the appetite of poultry.

Soy Alternatives

There has been interest in the last several years for rations that do not contain soy products. This presents a special challenge, as soy is very high in protein (36 to 38%), high in calories (1050 to 1600 Kcal/lb), high in fat (16 to 20%), and full of natural oils that provide non-starch energy. Soy meal is readily available, and functions as a cheap source of protein.

To make a soy-free ration you need to find a way to replace the protein. Sources that have been used include peas, sunflower meal, linseed (flax) meal, camelina meal, fish and crab meal. Each has its limitations.

Linseed Meal (Flax Meal)

Linseed meal is flax seeds that have had their oil removed by one of the methods described previously in the soybean section. Linseed meal is one of the better alternatives for replacing soy. Linseed meal will have approximately 30 to 37% protein, and 1100 to 1300 Kcal/lb energy. It is generally only used in ruminant feeds, but if kept fresh and not oxidized (rancid) it

Flax Seed

Poultry Feed Ingredients

also works well in poultry and swine feeds. Linseed may be fed up to 20% of the diet safely. Levels above 30% have a potential to add a fishy flavor or paint like smell to meat or eggs. Use caution when combining with fish meals and other high Omega-3 oils.

Sunflower Meal

Sunflower meal is a by-product of the oil seed industry. Typical protein levels are 34 to 38% and energy is 1000 to 1100 Kcal/lb. When fresh it can be used for up to 10 to 15% of the diet. However, sunflower meal is highly prone to oxidation and is very high in non-digestible fiber. If you choose to use sunflower meal you should remove any other feed ingredients that have high fiber content (oats, barley, alfalfa meal, etc.) in order to keep the overall diet fiber below 7 to 8%.

Sunflower Meal

I have found sunflower meal to not be a desirable ingredient in poultry rations. The birds do not eat it aggressively unless it is fresh.

Camelina Meal

Camelina meal is a by-product of the oil seed industry. Camelina is a member of the mustard family, currently being grown in areas with very low annual rainfall and marginal soil fertility, such as Montana. It was developed in Australia for bio-diesel as well as a feed stuff for animals. The seed contains 35 to 50% natural oil content. Camelina has an average protein content of 37%, with an energy content of 1510 Kcal/lb after being processed and the majority of the oil having been removed.

Camelina Seed

After the oil has been removed, camelina has equivalent protein and energy to roasted soybeans. Unfortunately, it is currently limited by the FDA and cannot exceed 10% of layer and broiler diets. This is a relatively new product, but so far does not appear to oxidize quickly.

Fundamental Ration Formulation

There are not many significant differences needed between diets for commercial poultry and pastured poultry. It is, however, very challenging to balance poultry rations around the pasture, as pasture changes with each season and with each region. Therefore, I recommend balancing a ration without including the nutritional contribution from pasture. When pasture is available it is a bonus nutritional input.

Feed Nutrient Values

The traditional feed nutrient values used for ration formulation are listed in Appendix C: Feed Ingredient Values and Spreadsheet Ration Calculator. These ingredient values are what I use in the formulation of poultry feed rations. They are accepted throughout the nutritional community as acceptable averages for commonly found feed ingredients. Appendix C shows the protein, fat, fiber, and energy values, as well as the macro mineral and vitamin values of numerous feedstuffs. I do not expect grains to contribute much vitamin value, since the vitamins in grains deplete quickly during storage. Many grains bought commercially today may be more than a year old and will have lost much of their vitamin content.

The last pages of Appendix C include the micro mineral and amino acid values. Usually 22 amino acid values are listed; I have listed only the four amino acids that are most critical for poultry. Whenever one is using a fairly conventional type ration the requirements for the four amino acids will be met. Whenever one or more of these four are deficient, I find that one or more of the other 22 will also be deficient. Therefore, over time I have concluded that once these four amino acid values have been met the others will be adequate.

Poultry Nutritional Requirements

The following appendices identify the commercial industry nutritional requirements for selected categories of poultry.
- Appendix D: Commercial Broiler-Roaster Nutritional Requirements
- Appendix E: Commercial Layer Nutritional Requirements
- Appendix F: Commercial Turkey Nutritional Requirements
- Appendix G: Commercial Meat Duck Nutritional Requirements

Fundamental Ration Formulation

These requirements are all based on commercial feeding recommendations. I have converted the values into U.S. measurements for easier interpretation. These pages identify the protein, energy, vitamin, and mineral requirements for each type of poultry.

Sample Commercial Rations

- Appendix H: Commercial Broiler Sample Rations
- Appendix I: Commercial Roaster Sample Rations
- Appendix J: Commercial Layer Sample Rations
- Appendix K: Commercial Turkey Sample Rations

The above information is based on confinement poultry in controlled climates, as identified by Leeson and Summers, in the book *Commercial Poultry Nutrition*, published in 1997. This information may or may not be suitable for pastured or ranged poultry operations. Pastured or range poultry do not have climate control. Therefore the nutritional requirements will be different based on geographical location.

Feeding Pasture-Raised Poultry

Sample Rations for Pastured Poultry

I have formulated rations that I feel are most suitable for various breeds and age stages of pastured poultry.

Pasture Broiler Sample Ration (Starter/Grower)

This chick starter-grower recipe is acceptable to use from day one until the desired body weight is achieved, typically between days 42 to 56. It may also be used for starting new replacement pullets from day 1 to day 56. After day 56 those raising laying hens should change to a 16% pullet developer ration.

Pasture Broiler Ration

Ingredients:	LBS
Alfalfa Meal	100
Aragonite	25
Corn Grain Shell	1015
Fish meal	75
Oats	100
Vitamin Mineral Premix	60
Soybeans, Roasted	625
Total	2000

Nutrient Name:	Units	Amount
Crude Protein	%	19.4%
Crude Fat	%	8.1%
Crude Fiber	%	4.5%
Calcium	%	1.30%
Phosphorus	%	0.79%
Salt added	%	0.34%
Sodium	%	0.17%
Energy	Kcal/LB	1,379
Vitamin A	IU/LB	4833
Vitamin D	IU/LB	1608
Vitamin E	IU/LB	50
Choline	IU/LB	476
Biotin	MCG/LB	50.7
Manganese	IU/LB	58.9
Zinc	IU/LB	47.9
Copper	IU/LB	4.03
Selenium (added)	IU/LB	0.30
Lysine	%	1.26%
Methionine	%	0.45%
Methionine/Cystine	%	0.64%
Arginine	%	1.25%

Sample Rations

Pasture Roaster Sample Ration

This is a recipe that you should choose to use only when you are raising meat production birds to final carcass weights above 5 pounds. Change to this formula from the starter ration on page 48 at the end of week six (day 42). Continue with this ration until the desired weight is achieved.

Pasture Roaster Ration

Ingredients:	LBS
Alfalfa Meal	100
Aragonite	25
Corn Grain Shell	1215
Fish meal	50
Oats	100
Vitamin Mineral Premix	60
Soybeans, Roasted	450
Total	2000

Nutrient Name:	Units	Amount
Crude Protein	%	16.1%
Crude Fat	%	6.8%
Crude Fiber	%	4.3%
Calcium	%	1.22%
Phosphorus	%	0.73%
Salt added	%	0.33%
Sodium	%	0.17%
Energy	Kcal/LB	1,384
Vitamin A	IU/LB	4908
Vitamin D	IU/LB	1608
Vitamin E	IU/LB	50
Choline	IU/LB	479
Biotin	MCG/LB	54.0
Manganese	IU/LB	58.9
Zinc	IU/LB	47.7
Copper	IU/LB	4.10
Selenium (added)	IU/LB	0.30
Lysine	%	1.05%
Methionine	%	0.40%
Methionine/Cystine	%	0.54%
Arginine	%	1.01%

Feeding Pasture-Raised Poultry

Pasture Layer Sample Ration, First Laying Cycle
This recipe is well designed for layers in their first laying cycle that do not have controlled, limited feed access. Raise to pullets on the starter ration (page 48) and then switch to this layer ration when the pullet has reached 18 weeks of age. This recipe may be fed for the entire first laying cycle, usually 50 to 60 weeks of egg production.

Pasture Layer First Cycle Ration

Ingredients:	LBS
Alfalfa Meal	100
Aragonite	175
Corn Grain Shell	965
Oats	100
Vitamin Mineral Premix	60
Soybeans, Roasted	600
Total	2000

Nutrient Name:	Units	Amount
Crude Protein	%	16.5%
Crude Fat	%	7.5%
Crude Fiber	%	4.3%
Calcium	%	3.89%
Phosphorus	%	0.74%
Salt added	%	0.30%
Sodium	%	0.17%
Energy	Kcal/LB	1,274
Vitamin A	IU/LB	4814
Vitamin D	IU/LB	1608
Vitamin E	IU/LB	50
Choline	IU/LB	455
Biotin	MCG/LB	48.7
Manganese	IU/LB	69.3
Zinc	IU/LB	46.4
Copper	IU/LB	4.28
Selenium (added)	IU/LB	0.30
Lysine	%	1.04%
Methionine	%	0.37%
Methionine/Cystine	%	0.54%
Arginine	%	1.09%

Sample Rations

Pasture Layer Sample Ration Second Laying Cycle
This recipe is best used for laying hens that have completed their first laying cycle and are beginning their second. Since they are more mature and with more body weight second-cycle hens will require more feed than hens during their first cycle. With the natural increase of feed consumption it is necessary to reduce the protein level to help keep egg size in the desired range. Excessive protein will cause more jumbo or super jumbo eggs.

Pasture Layer Second Cycle Ration

Ingredients:	LBS
Alfalfa Meal	100
Aragonite	200
Corn Grain Shell	1040
Oats	50
Vitamin Mineral Premix	60
Soybeans, Roasted	550
Total	2000

Nutrient Name:	Units	Amount
Crude Protein	%	15.6%
Crude Fat	%	7.0%
Crude Fiber	%	4.2%
Calcium	%	4.34%
Phosphorus	%	0.74%
Salt added	%	0.30%
Sodium	%	0.17%
Energy	Kcal/LB	1,266
Vitamin A	IU/LB	4842
Vitamin D	IU/LB	1608
Vitamin E	IU/LB	50
Choline	IU/LB	457
Biotin	MCG/LB	49.5
Manganese	IU/LB	70.8
Zinc	IU/LB	46.5
Copper	IU/LB	4.25
Selenium (added)	IU/LB	0.30
Lysine	%	0.99%
Methionine	%	0.36%
Methionine/Cystine	%	0.50%
Arginine	%	1.01%

Feeding Pasture-Raised Poultry

Pasture Turkey Sample Starter Ration

The turkey starter recipe is best used from day 1 through day 28. It is very important to offer young turkeys a higher nutrient density when starting than that which you offer broilers. Turkeys do not learn to eat as quickly as broilers do, so you need to make every mouthful count. You may want to spend extra time with young turkeys during the first 3 days showing them where the feed and water are.

Pasture Turkey Starter Ration

Ingredients:	LBS
Alfalfa Meal	100
Aragonite	25
Poultry Nutri-Balancer	80
Soybeans, Roasted	880
Sea-Lac, Fish meal	100
Shell Corn Grain	615
Wheat	200
Total	2000

Nutrient Name:	Units	Amount
%DM	%	90%
%CP	%	24.2%
%CF	%	10.3%
%C Fiber	%	4.9%
%CAL	%	1.54%
%PHO	%	0.98%
%Salt add	%	0.45%
%Sodium	%	0.21%
ME/Kcal/lb	Kcal/LB	1,361
Vit A	IU/LB	6179
Vit D	IU/LB	2144
Vit E(add)	IU/LB	67
Choline	IU/LB	559
Biotin	MCG/LB	52.7
Manganese	IU/LB	73.9
Zinc	IU/LB	58.2
Copper	IU/LB	4.79
Selenium (added)	IU/LB	0.40
Lysine	%	1.52%
Methionine	%	0.54%

Sample Rations

Pasture Turkey Sample Grower #1 Ration

Turkey grower #1 ration should be used from day 29 through day 56. This period is very important for proper organ and immune system development. This is also the period where most of the bone structure is developed.

Pasture Turkey Grower #1 Ration

Ingredients:	LBS
Alfalfa Meal	50
Aragonite	25
Poultry Nutri-Balancer	80
Soybeans, Roasted	750
Sea-Lac, Fish meal	100
Shell Corn Grain	795
Wheat	200
Total	2000

Nutrient Name:	Units	Amount
%DM	%	89%
%CP	%	22.1%
%CF	%	9.3%
%C Fiber	%	4.2%
%CAL	%	1.45%
%PHO	%	0.96%
%Salt add	%	0.45%
%Sodium	%	0.22%
ME/Kcal/lb	Kcal/LB	1,382
Vit A	IU/LB	6246
Vit D	IU/LB	2144
Vit E(add)	IU/LB	67
Choline	IU/LB	577
Biotin	MCG/LB	56.0
Manganese	IU/LB	74.2
Zinc	IU/LB	58.9
Copper	IU/LB	4.91
Selenium (added)	IU/LB	0.40
Lysine	%	1.39%
Methionine	%	0.51%

Feeding Pasture-Raised Poultry

Pasture Turkey Sample Grower #2 Ration

Turkey Grower #2 ration is used from days 57 to 85. This is when the muscle is formed and the structure further developed.

Pasture Turkey Grower #2 Ration

Ingredients:	LBS
Alfalfa Meal	100
Aragonite	25
Poultry Nutri-Balancer	60
Soybeans, Roasted	600
Fish meal, 60%	100
Shell Corn Grain	915
Wheat	200
Total	2000

Nutrient Name:	Units	Amount
%DM	%	89%
%CP	%	20.1%
%CF	%	8.3%
%C Fiber	%	4.6%
%CAL	%	1.36%
%PHO	%	0.83%
%Salt add	%	0.35%
%Sodium	%	0.17%
ME/Kcal/lb	Kcal/LB	1,375
Vit A	IU/LB	4807
Vit D	IU/LB	1608
Vit E(add)	IU/LB	50
Choline	IU/LB	528
Biotin	MCG/LB	48.1
Manganese	IU/LB	59.3
Zinc	IU/LB	48.3
Copper	IU/LB	4.08
Selenium (added)	IU/LB	0.30
Lysine	%	1.26%
Methionine	%	0.47%

Sample Rations

Pasture Turkey Sample Finisher Ration

Turkey Finisher Ration is best used from day 86 through desired finished weights. This lower protein formula allows for proper body fat deposits, ensuring great flavor and tenderness.

Pasture Turkey Finisher Ration

Ingredients:	LBS
Alfalfa Meal	100
Aragonite	25
Poultry Nutri-Balancer	60
Soybeans, Roasted	450
Sea-Lac, Fish meal	50
Shell Corn Grain	915
Wheat	400
Total	2000

Nutrient Name:	Units	Amount
%DM	%	89%
%CP	%	17.0%
%CF	%	7.3%
%C Fiber	%	4.7%
%CAL	%	1.23%
%PHO	%	0.75%
%Salt add	%	0.33%
%Sodium	%	0.17%
ME/Kcal/lb	Kcal/LB	1,368
Vit A	IU/LB	4820
Vit D	IU/LB	1608
Vit E(add)	IU/LB	50
Choline	IU/LB	562
Biotin	MCG/LB	47.4
Manganese	IU/LB	60.2
Zinc	IU/LB	47.4
Copper	IU/LB	4.32
Selenium (added)	IU/LB	0.30
Lysine	%	1.00%
Methionine	%	0.41%

Feeding Pasture-Raised Poultry

Soy-Free Rations
Those wishing to feed rations without soy may try these recipes. See the research on soy-free alternatives in Appendix B.

Broiler Ration No Soy

Ingredients	LBS
Aragonite	25
Camelina	200
Corn, medium ground	690
Crab Meal	150
Fish Meal 64%	100
Linseed (Flax) Meal	300
Peas, field	200
Poultry Nutri-Balancer	60
Wheat, 12%	200
Total	2000

Nutriet Name	Units	Amount
Crude Protein	%	19.8%
Crude Fat	%	7.2%
Crude Fiber	%	5.2%
Calcium	%	2.38%
Phosphorus	%	1.00%
Salt added	%	0.70%
Sodium	%	0.26%
Energy	KCAL/LB	1.235
Vitamin A	IU/LB	4723
Vitamin D	IU/LB	1608
Vitamin E	IU/LB	50
Choline	IU/LB	1521
Biotin	MCG/LB	45.2
Manganese	IU/LB	147.8
Zinc	IU/LB	93.2
Copper	IU/LB	15.86
Selenium	IU/LB	0.30
Lysine	%	0.88%
Methionine	%	0.49%
Methonine/Cystine	%	0.67%
Arginine	%	1.26%

Sample Rations

Layer Ration No Soy

Ingredients	LBS
Aragonite	150
Camelina	200
Corn, medium ground	790
Crab Meal	100
Fish Meal 64%	50
Linseed (Flax) Meal	200
Oil, Vegetable	50
Peas, field	200
Poultry Nutri-Balancer	60
Wheat, 12%	200
Total	2000

Nutriet Name	Units	Amount
Crude Protein	%	16.2%
Crude Fat	%	5.7%
Crude Fiber	%	4.7%
Calcium	%	4.24%
Phosphorus	%	0.91%
Salt added	%	0.59%
Sodium	%	0.23%
Energy	KCAL/LB	1177
Vitamin A	IU/LB	4761
Vitamin D	IU/LB	1608
Vitamin E	IU/LB	50
Choline	IU/LB	1521
Biotin	MCG/LB	45.9
Manganese	IU/LB	147.5
Zinc	IU/LB	89.3
Copper	IU/LB	14.32
Selenium	IU/LB	0.30
Lysine	%	0.67%
Methionine	%	0.40%
Methonine/Cystine	%	0.55%
Arginine	%	1.04%

Feeding Pasture-Raised Poultry

Alternative Rations

Most of you probably raise Cornish cross broilers, or some type of hybrid hen. The chickens that most of us raise thrive on corn- and soy-based diets because their parents, the breeder flocks, have been selected to thrive on corn- and soy-based diets. We can't get too far from that base line of nutrition or our birds will not thrive. This limitation allows about 25 to 30% of the diet to be alternative ingredients, some of the time.

The people in charge of breeding flocks producing the eggs from which your chickens are hatched feed least-cost rations. These rations are made primarily from corn and soy. As soon as the feed manufacturers reach the basic diet requirements for the class of poultry for which they are mixing, they add fillers such as bakery by-product waste meal, dried corn distillers, salvaged grain dust, etc. I really don't even want to know what all goes into those rations. These ingredients are used to lower the price of the feed in circumstances where $0.50 per ton will save a mega company like Tyson foods over $10,000 per day!

You can't imagine what I have been asked to put into a chicken ration. I have been asked to make rations using organically grown ground beef, nutmeg pressing residue, coconut meal, soldier fly larvae, whole milk, skim milk, dried milk, clabbered milk, crickets, fish offal, wheat sprouts, ethanol dried distillers grains, etc. etc. A chicken can be made to eat any or all of these things. A laying hen will eat some of the most disgusting things you can think of, and that on her own free will: road kill, any dead carcass, scratched open manure patties, small rodents and almost any insect.

Most of the time I embrace the challenge and learning curve of trying to make specialty things fit into a poultry diet. But, we need to remember that the nutrition (food ??) she is eating is (ultimately the food??) the nutrition we all are eating. We're getting it in the egg or in the meat of a broiler. I recommend that we choose more wisely what we put in our birds' feed rations. You are going to eat some of those chickens, aren't you? Personally I like the chickens that I eat to get a few oats, barley, wheat and triticale when they are in season–good, whole grain nutrition, not someone's leftovers, offal or by-products.

At a small scale, skimping on feed ingredients won't save a significant enough amount of money to really make a difference over the long run. For example, a ton of feed is mathematically enough to raise 133 Cornish cross broilers to 5 pound carcass weights. If you save $5.00 per ton of broiler feed, it will only make

Alternative Rations

a difference of 3.759¢ per bird. In the case of layer feed, a savings of $5.00 per ton of feed (a ton contains 7,407 feed servings for a hen, and will feed 200 hens for 37 days) equals an increase in cost of only 1.013514¢ per dozen eggs. If you are squeezing the nickels that hard, I suggest that you QUIT raising chickens! If you aren't able to raise 133 Cornish cross broilers on a ton of feed, you either need to get better feed or improve your poultry management and improve the poultry's living conditions.

Until a breed of chicken is developed to thrive on milk, milk by-products, coconut meal, nutmeg extract residue and all the other possibilities that I mentioned earlier, let's stay relatively close to the basics with our feed ingredients. I will keep trying to develop rations using the items that some of you request. Sometimes I will smile when you ask, sometimes I will just question why, and sometimes I will disagree with you. But please remember, regardless of our conversation or correspondence, my job is to help you be successful! So before you ask me to add some Fu-Fu dust, snake oil, or strange cheap by-product, make sure you would eat it yourself before giving it to your chickens (or any of your other animals for that matter). If you put garbage in you get garbage out!

Feeding Pasture-Raised Poultry

Formulating Your Own Rations

Those of you with access to your own grains, or who have grains available from neighbors or locally may want to try developing your own ration. As poultry have very specific nutritional needs, especially when in the fast growing stage, you must pay very close attention to the nutritional content of each ingredient as well as the resulting finished product.

Ration-Balancing By Hand

Refer to Appendix L: Formulating Rations with the Pearson Square.

The Pearson square, or box method of balancing rations, is a simple procedure that has been used for many years. It is of greatest value when only two ingredients are mixed. I have included a description and instructions on how to use the Pearson Square in Appendix L.

Ration-Balancing on a Computer Spreadsheet

Refer to Appendix C: Feed Ingredient Values and Spreadsheet Ration Calculator.

I designed the spreadsheet included in Appendix C so that you can put together your own rations.

Directions for use:
1. There are columns provided between each nutritional component for calculations.
2. Multiply the desired pounds of each feedstuff with corresponding component value. Place that value in the right side column.
3. When all of the feedstuffs have been calculated, add each column of calculated values vertically downward to arrive with a total at the bottom of each column in the totals row.
4. Divide each total value by the total weight of the desired amount of feed to determine the component level of the proposed ration.
5. Compare total values with required or desired values and make adjustments as needed by modifying quantities of ingredients.
6. Make copies of these spreadsheets prior to making calculations and markings on the original (master) spreadsheet.

Feeding Ducks

Ducks have been bred and managed in much the way as Cornish cross broiler and commercial layers, and so their dietary requirements are very close to other commercial poultry. Anyone currently raising pastured broilers and layers can use existing poultry feeds "as is," or slightly modified (see below) for raising duck for meat and/or eggs.

If your ducklings appear weak, lethargic or stunted, add 1.2 pound of fishmeal to 30 pounds (5 gal bucket) of broiler feed. This will boost the protein from 19.5% to 22%, and will help those birds. This is not necessary for normal healthy ducklings, but is helpful if you detect a weakness or deficiency. This addition should only be required for 2 to 3 weeks, after which you can transition the ducklings to a grower diet. During the grower stage of development for ducks, (typically starting at week 4 till the end of week 8), a feed protein content lower than 19.5% is desirable. This can be achieved by offering an additional feeder with 40% cracked corn, 40% whole wheat or barley, and 20% grower grit (measured by weight) and allowing the birds to free choice from that feeder.

Pasture Duck Sample Rations

Duck Starter 22.5% Protein Day 1 to 21		Duck Grower 17.7% Protein Day 22 to Finish	
Ingredients:	LBS	Ingredients:	LBS
Alfalfa Meal	100	Alfalfa Meal	100
Aragonite	25	Aragonite	25
Fish meal,64%	50	Poultry Nutri-Balancer	60
Poultry Nutri-Balancer	60	Shell Corn Grain, cracked	915
Shell Corn Grain medium grind	815	Soy Bean Meal,48%	250
Soy Bean Meal,48%	400	Soybeans, Roasted	250
Soybeans, Roasted	300	Wheat, Whole	400
Wheat,cracked	250	Total	2000
Total	2000		

Feeding Pasture-Raised Poultry

Pastured Duck Layers

Duck layer feed requirements are very similar to a chicken layer diet–requiring adequate protein for good production and egg size as well as dietary calcium for proper shell formation and other mineral uptake and absorption. There is nothing unusually specific for a duck layer diet. However, since they like to forage, if you feed scratch grains their diet is often neglected. You can feed this way, but it doesn't maximize the laying potential. If your market is to sell eggs, feed the laying ducks the following for optimum production. If they are just cute creatures to decorate your landscape just feed them scratch grains.

Pastured Duck Layer Ration   16.9% Protein

Ingredients:	LBS
Alfalfa Meal	100
Aragonite	175
Corn Grain Shell	940
Oats	100
Poultry Nutri-Balancer	60
Soybeans, Roasted	625
Total	2000

During the off-season laying ducks should be fed a holding diet of 13% protein. This is the perfect place to feed a scratch grain blend of cracked corn, whole oats and whole wheat along with grit, oyster shells and 3% Poultry Nutri-Balancer. Thirty days before the onset of the new laying cycle the layers should be changed back to the production laying diet to increase blood calcium levels as well as vitamins and critical trace minerals prior to egg formation and lay.

Feeding Recommendations for Meat Ducks and Young Layers

	Starter #1	Starter #2	Grower #1	Grower #2
Approximate Protein %	22	20	18	16
Met. Energy (Kcal/Lb)	1295	1320	1400	1420
Calcium (%)	0.8	0.83	0.76	0.75
Available Phosphorus (%)	0.4	0.42	0.38	0.35

Folic Acid (mg)	0.23	Vitamins (per Lb)		
Biotin (mg)	0.1	Vitamin A (I.U.)	3650	
Niacin (mg)	28	Vitamin D (I.U.)	1140	
Vitamin K (mg)	0.7	Choline (mg)	365	
Vitamin E (I.U.)	10	Riboflavin (mg)	1.8	
Thiamin (mg)	1	Pantothetic Acid (mg)	5.5	
Pyridoxine (mg)	1.4	Vitamin B12 (mg)	0.005	

* from *Commercial Poultry Nutrition*, 2nd edition, by Leeson & Summers

Multi-Species Feed Ration

I originally designed this multi-species ration with 16% protein for a few farmers in the New England states who were keeping just a few head of many different animals on the same farm. This was a common practice a couple of generations ago. I can remember my grandfather keeping two milk cows, half a dozen pigs, 20 to 50 chickens (depending on the season), a couple of sheep to eat the hedge rows and rough areas, a pony or two for the grandkids, etc. You really never knew what you might see next at Grandpa's place.

There are hundreds of small self-reliant farms around the countryside where farmers are still raising a little of this and a few of that. In this situation, it is not realistic to make custom feeds specific for each animal's needs. However, many small farmers see many reasons to get away from the commercial feed industry's products: GMOs, by-products, and animal residues to name some of the more common ones.

This multi-species ration is generic, one-size-fits-all. I want to be very clear that this formula is not specific to any one animal or poultry type. It will work well for all of the species that most folks will keep on their farms. But it will not provide "perfect" diet requirements as prescribed by the commercial feed industry. The animals will all perform at a slower rate of gain or production, but they will be healthy and happy. I don't want anyone to think that their animals will perform as well on this multi-species ration as on those specific to each species.

Base Ration Formulations

Soy Multi-Species Base Ration

Base Ration 16% Protein

Ingredients	LBS	LBS	Protein
Corn, course ground	1050	262.5	8%
Oats, crimped/ ground/whole	300	75	10%
Roasted soy	650	162.5	38%
Totals	2000	500	18.05%

Feeding Pasture-Raised Poultry

No Soy Multi-Species Base Ration

Base Ration 16% Protein

Ingredients	LBS	LBS	Protein
Corn, course ground	1000	250	9%
Oats, crimped/ ground/ whole	250	62.5	10%
Flax meal	250	62.5	38%
Sunflower meal	300	75	34%
Triticale	200	50	12%
Totals	2000	500	16.8%

All poultry should have free choice grit at all times. All non-poultry livestock should have Fertrell's Grazier's Choice offered free choice at all times.

Animal-Specific Additions

The additives and amounts listed for each animal are based on 30 pounds (a 5-gallon pail) of base ration feed. For each animal listed, add the prescribed amounts of each additive to the base ration.

Layers
Poultry Nutri-Balancer .... 1 lb
Aragonite (Calcium) ....2.75 lbs

Broiler Chicks
Poultry Nutri-Balancer .... 1 lb
Aragonite (Calcium) .... 1 lb
Fish Meal .... 1.5 lbs

Turkeys, 1-4 weeks
Poultry Nutri-Balancer ....1.5 lbs
Aragonite (Calcium) .... 0.375 lbs
Fish Meal .... 4 lbs
Roasted Soy .... 4 lbs

Turkeys, 4-8 weeks
Poultry Nutri-Balancer .... 1 lb
Aragonite (Calcium) .... 0.375 lbs
Fish Meal .... 2 lbs

Multi-Species Rations

Turkeys after week 8 until finished
Whole Wheat..........................Free Choice
Mixed with grit 4 parts wheat to 1 part grit

Cattle
Dairy Nutri-Balancer #3 .... 1.5lbs
Aragonite (Calcium)............. 1 lb
Rumi Cult 40 .... 1.5 lbs

Goats
Goat Nutri-Balancer .... 2 lbs
Rumi Cult 40 .... 1 lb

Horses
Horse Power .... 2 lbs
Rumi Cult 40 .... 1 lb

Pigs
(no soy ration)
Swine Growers Premix .... 1 lb

Feeding Pasture-Raised Poultry

Laying Hen Considerations

Pastured poultry producers have different reasons for maintaining a layer flock. Some want hens as cattle manure managers, or as insect control technicians, or as fluffy feathery friends (pets or farm decoration). Others raise hens to provide an additional source of farm income. Each of these is justifiable, but each need different styles of management to achieve the desired end. The following thoughts will assist you in determining which type of flock is best for you.

Determine your primary purpose in managing a laying flock:
-Adding income to the overall farming operation. Research at farmers markets, food clubs or co-ops, or local grocery stores should give you a good indication of what price you could sell your eggs for. Your ability to sell at these various markets will depend on your state and local laws for farm egg marketing
-Land or insect management. If the primary purpose of keeping hens is for redistributing cattle manure droppings and eating insects, they will generally not be a significant income maker, or even a break-even farm enterprise.
-Keeping a small flock of hens around to just make the yard look good and have enough eggs for the family and friends.

After you decide which production category you would like to be in, make the correct selections of breed, housing, environmental controls, and feed program to achieve your goal.

High production varieties are capable of 275 to 300 eggs per first cycle of lay. Old fashioned varieties typically produce 225 to 250 eggs per first cycle of lay.

Flocks for maximum profit are:

• High production breed varieties such as Bovan Brown, ISA Goldens, Golden Comets, Red Stars, all mostly Red Sexlink varieties.
• These birds must be fed a 17% to 18% protein peaking layer feed that is properly balanced to maximize the breed's potential, adjusted for each stage of the lay cycle.
• The birds must be properly stimulated with the use of an artificial lighting schedule starting at 12 hours, going up in increments of no more than 30 minute

Laying Hen Considerations

per day to 15.5 to 16 hours, and then settling back to 14 to 14.5 hours to ensure they reach maximum peak production.
• A clean, dry, fresh air environment must be maintained with plenty of access to fresh clean water, feed, grit and coarse calcium such as oyster shell.
• Each hen will eat on average 0.27 pounds of feed per day over the course of the lay cycle, or 1 year. This is 98.55 pounds per year per hen. You should limit the feed being consumed each day to minimize overweight hens, reduce oversized eggs, and maximize production. Controlled feeding requires enough feeder space for all hens to eat at the same time.
• Split feedings of twice or more per day are very beneficial. I recommend feeding 40% in the morning, and the remaining 60% about an hour before sunset. You should also consider a three-phase feeding program, 18% protein phase 1, 16.5% protein phase 2 and 15% protein phase 3
• A reasonable expectation for this type of flock is 290 or more eggs per chicken in a 52-week lay cycle.

Flocks used as cattle manure managers, insect control technicians or fluffy feathery friends are:

• Older breed varieties such as Buff Orpingtons, Black Australorp, Barred Rocks, Rhode Island Reds, and so on.
• Fed a feed from 12% to 16% protein that is somewhat balanced but not precisely for one specific breed of poultry or fowl.
• Do not need supplemental light for stimulation and are allowed to molt during the winter months.
• May be housed in almost any structure, from cow stables to greenhouses.
• Feed can be filled once or twice a week using a bulk-type feeder, and hens may be required to find their own grit and coarse calcium.
• Each hen will eat on average 0.35 pounds of feed each day over the course of the lay cycle, or 1 year. This is 127.75 pounds per year per hen.
• This flock doesn't have any specific production expectations. With good management they could provide up to 255 eggs per hen per 52-week lay cycle.

Let's compare the bottom line. If feed costs $400 per ton, it is 20¢ per pound.

Production layers produce approximately 300 eggs per year, are fed 98.55 lbs of feed at .20/lb = $19.71 of feed. This calculates to 6.57¢ in feed cost for each egg produced.
Heritage breed layers produce 255 eggs per year and will eat 127.71 lbs of feed which will cost you $25.55. This comes to 10¢ in feed cost for each egg produced.

For many years I happily maintained a "yard decoration" flock in my backyard that I didn't make money on. But I don't make my living from selling eggs or being a farmer, so it was ok to consider it a hobby and just enjoy my own test flock for learning purposes and eggs.

Diagnose and Treat Common Illnesses

Over the past 16 years of working closely with pastured poultry producers, organic commercial growers and hobbypoultry enthusiasts, I have listened to, observed and monitored many illness symptoms, living conditions, breed characteristics, and environmental influences. Observations of numerous illnesses inspired hours of great conversation with my mentor, Jack A. Robinette.

These conversations helped me identify the illnesses, based on as many symptoms as I could gather. Jack would confirm what a disease was, and explain how the conventional poultry system would treat it.

After Jack identified treatments, I had to go find a natural or organic treatment alternative. I really think that Jack enjoyed the challenge and mental exercise of trying to help me find alternative, natural and organic-acceptable treatment therapies.

The following flowchart is a culmination of my learning and observations, with a huge input of help from Jack Robinette.

This flow chart does not include all possibilities, merely the most common. These remedies are not mainstream commercial treatments. They have worked for others but they may not work all of the time. It is my goal that this information helps some of you some of the time.

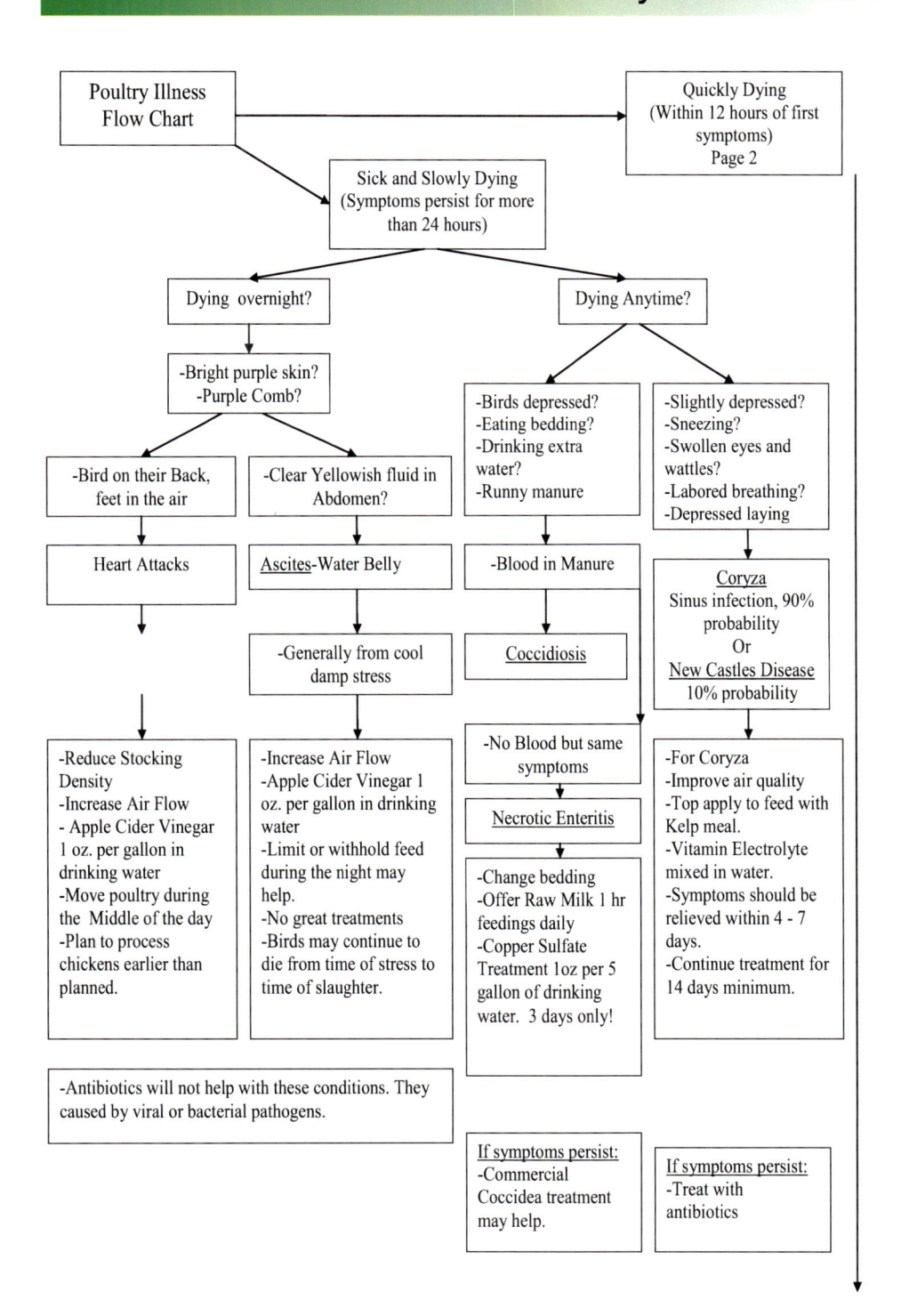

Poultry Illness Flow Chart

Poultry Illness Flow Chart → **Quickly Dying** (Within 12 hours of first symptoms) Page 2

↓

Sick and Slowly Dying (Symptoms persist for more than 24 hours)

Dying overnight?

- Bright purple skin?
- Purple Comb?

Bird on their Back, feet in the air
→ **Heart Attacks**

- Reduce Stocking Density
- Increase Air Flow
- Apple Cider Vinegar 1 oz. per gallon in drinking water
- Move poultry during the Middle of the day
- Plan to process chickens earlier than planned.

Clear Yellowish fluid in Abdomen?
→ **Ascites** - Water Belly
- Generally from cool damp stress

- Increase Air Flow
- Apple Cider Vinegar 1 oz. per gallon in drinking water
- Limit or withhold feed during the night may help.
- No great treatments
- Birds may continue to die from time of stress to time of slaughter.

- Antibiotics will not help with these conditions. They caused by viral or bacterial pathogens.

Dying Anytime?

Birds depressed? Eating bedding? Drinking extra water? Runny manure
→ **Blood in Manure** → **Coccidiosis**

- No Blood but same symptoms → **Necrotic Enteritis**

- Change bedding
- Offer Raw Milk 1 hr feedings daily
- Copper Sulfate Treatment 1oz per 5 gallon of drinking water. 3 days only!

If symptoms persist:
- Commercial Coccidea treatment may help.

Slightly depressed? Sneezing? Swollen eyes and wattles? Labored breathing? Depressed laying

Coryza Sinus infection, 90% probability
Or
New Castles Disease 10% probability

- For Coryza
- Improve air quality
- Top apply to feed with Kelp meal.
- Vitamin Electrolyte mixed in water.
- Symptoms should be relieved within 4 - 7 days.
- Continue treatment for 14 days minimum.

If symptoms persist:
- Treat with antibiotics

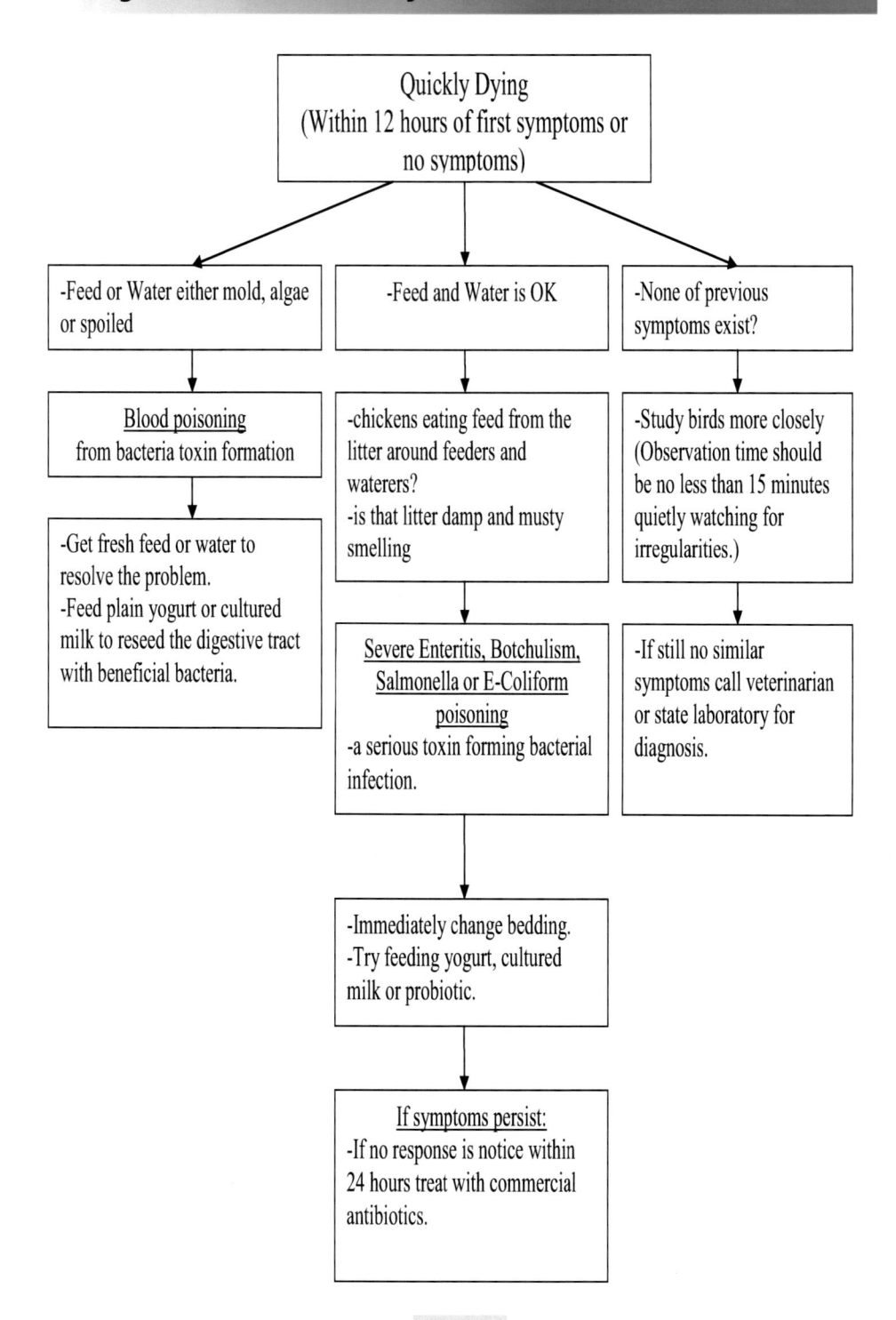

Feeding Pasture-Raised Poultry

Quickly Dying
(Within 12 hours of first symptoms or no symptoms)

- Feed or Water either mold, algae or spoiled

Blood poisoning
from bacteria toxin formation

- Get fresh feed or water to resolve the problem.
- Feed plain yogurt or cultured milk to reseed the digestive tract with beneficial bacteria.

- Feed and Water is OK

- chickens eating feed from the litter around feeders and waterers?
- is that litter damp and musty smelling

Severe Enteritis, Botchulism, Salmonella or E-Coliform poisoning
- a serious toxin forming bacterial infection.

- Immediately change bedding.
- Try feeding yogurt, cultured milk or probiotic.

If symptoms persist:
- If no response is notice within 24 hours treat with commercial antibiotics.

- None of previous symptoms exist?

- Study birds more closely (Observation time should be no less than 15 minutes quietly watching for irregularities.)

- If still no similar symptoms call veterinarian or state laboratory for diagnosis.

Leg Problem Flow Chart

Leg Problem Flow Chart

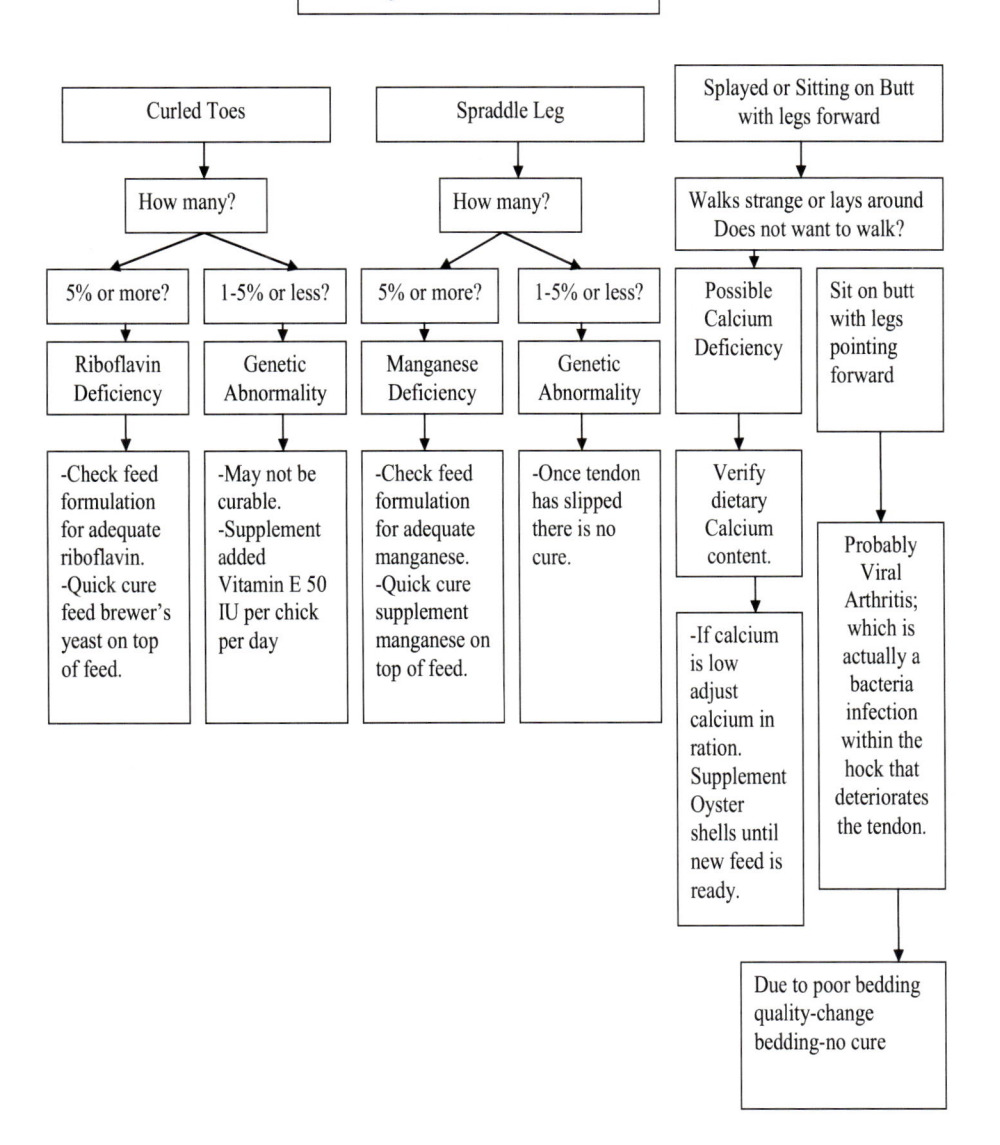

- **Curled Toes** → How many?
  - **5% or more?** → Riboflavin Deficiency
    - Check feed formulation for adequate riboflavin.
    - Quick cure feed brewer's yeast on top of feed.
  - **1-5% or less?** → Genetic Abnormality
    - May not be curable.
    - Supplement added Vitamin E 50 IU per chick per day

- **Spraddle Leg** → How many?
  - **5% or more?** → Manganese Deficiency
    - Check feed formulation for adequate manganese.
    - Quick cure supplement manganese on top of feed.
  - **1-5% or less?** → Genetic Abnormality
    - Once tendon has slipped there is no cure.

- **Splayed or Sitting on Butt with legs forward** → Walks strange or lays around. Does not want to walk?
  - **Possible Calcium Deficiency**
    - Verify dietary Calcium content.
    - If calcium is low adjust calcium in ration. Supplement Oyster shells until new feed is ready.
  - **Sit on butt with legs pointing forward**
    - Probably Viral Arthritis; which is actually a bacteria infection within the hock that deteriorates the tendon.
    - Due to poor bedding quality-change bedding-no cure

Note: This flow chart does not include all possibilities merely the most common. These remedies are not mainstream commercial treatment. They have worked for others but they may not work all of the time. It is my wish that it helps some of you some of the time.

Jeff Mattocks

Feeding Pasture-Raised Poultry

Appendices

Page	
73	**Appendix A: Poultry Nutri-Balancer**
75	**Appendix B: No-Soy Ration Research**
90	**Appendix C: Feed Ingredient Values and Ration Calculator**
95	**Appendix D: Commercial Broiler-Roaster Nutritional Requirements**
96	**Appendix E: Commercial Layer Nutritional Requirements**
97	**Appendix F: Commercial Turkey Nutritional Requirements**
98	**Appendix G: Commercial Meat Duck Nutritional Requirements**
99	**Appendix H: Commercial Broiler Sample Rations**
100	**Appendix I: Commercial Roaster Sample Rations**
101	**Appendix J: Commercial Layer Sample Rations**
102	**Appendix K: Commercial Turkey Sample Rations**
103	**Appendix L: Pearson Square**

Appendix A: Poultry Nutri-Balancer Improves Bird Livability and Mortality

By Jody Padgham
From *Fearless Farm Finances: Financial Management Demystified*, published by the Midwest Organic and Sustainable Education Service (MOSES) 2012

Planning to increase production, raising profitability was a goal at Rosebank Farms. Andrea Gunner identified that death losses in their roasting chicken production were costing them a lot. She read about a mineral supplement for poultry, Fertrell's Poultry Nutri-Balancer (PNB), that other producers claimed significantly reduced thriftyness-related losses and decreased feed costs by enhancing feed efficiency. The PNB, produced in central Pennsylvania, would cost Andrea $125 plus $10 shipping per 50-pound bag. The question she wanted to answer was: Will the positive effects of the Poultry Nutri-Balancer offset the costs of bringing it to British Columbia, on the west coast of Canada?

To answer her question, Andrea did a trial run of broilers using the PNB to see how much it affected mortality and feed consumption. The first two batches of birds (which, because of weather, traditionally have higher mortality), experienced losses of 19 and 22 percent without the mineral. The third batch, using Poultry Nutri-Balancer, experienced only 1.5 percent mortality, and subsequent batches had no measurable mortality at all. Previous records kept on Excel spreadsheets told Andrea that average mortality for an entire season without Poultry Nutri-Balancer was 16.5 percent. In the trial year, in which the mineral mix was used for all but the first two batches, the overall annual mortality was reduced to 7.4 percent. She also found that when fed the mineral, each chicken ate measurably less feed.

Now that Andrea could see that the Nutri-Balancer had an impact, she needed to find out if the financial impact of lower losses and lower feed use balanced out the increased cost of the supplement. Again using an Excel spreadsheet, and taking average costs and mortality figures from previous years, Andrea was able to figure that providing the mineral cost 17 cents per bird, but that the feed costs went down, from $4.70 to $3.94 per bird. Since all other costs were equal, this gave her a total net cost to produce birds with the PNB of $10.24 per bird, and $10.82 without. This alone might have convinced Andrea that this was a worthwhile change, but she also took into account the financial impact of the lower mortalities. This made the difference even more significant, with birds given the supplement costing only $10.50 per bird to produce, and those without costing her $11.41. See the chart detailing what Andrea learned on page 74.

Given that Andrea didn't need to make any other changes to use the Poultry Nutri-Balancer, she was able to make the easy decision to import the 50-pound

Feeding Pasture-Raised Poultry

bags of mineral from Fertrell in Pennsylvania, and actually save money while she was still charging the same price to her customers. Doing the trial run and tracking the real numbers gave her great confidence in making this decision, which at first glance may have looked expensive. If any part of her carefully calculated costs change over time, she will be able to easily enter new numbers and see what production or pricing changes she needs to make to accommodate them.

With Poultry Mineral Mix				Without Poultry Mineral mix			
2010		Chicken Costs		2010		Chicken Costs	
(based on 2,000)		$/MT	$/bird	(based on 2,000)		$/MT	$/bird
Chicks (incl freight)			1.74	Chicks (incl freight)			1.74
Apple cider vinegar			0.02	Apple cider vinegar			0.02
Skim milk powder			0.06	Skim milk powder			0.06
Fertrell's Poultry Nutri-Balancer			0.17	Fertrell's Poultry Nutri-Balancer			0.00
Grit		66.00	0.03	Grit		66.00	0.03
Starter	0.0011814	450.00	0.53	Starter	0.001425	450.00	0.64
Grower	0.00874884	390.00	3.41	Grower	0.0104125	390.00	4.06
Cracked Grain	0	400.00	0.00	Cracked Grain	0	400.00	0.00
Slaughter			3.25	Slaughter			3.25
Bags, Labels & Ties			0.15	Bags, Labels & Ties			0.15
H2O2 (4l), 35%, want 2500		42.99	0.01	H2O2 (4l), 35%, want 2500		42.99	0.01
Thymol (15 ml)		7.99	0.00	Thymol (15 ml)		7.99	0.00
Heat	350		0.18	Heat	350		0.18
Bedding	0.67	375.00	0.13	Bedding	0.67	375.00	0.13
Supplies	100		0.05	Supplies	100		0.05
Building R & M	400		0.20	Building R & M	400		0.20
Fencing R & M	100		0.05	Fencing R & M	100		0.05
Vehicle	500		0.25	Vehicle	500		0.25
Sub-Total Direct Costs			10.24	Sub-Total Direct Costs			10.82
Mortality	0.0741	3.44	0.25	Mortality	0.165	3.55	0.59
Total Direct Costs			**10.50**	**Total Direct Costs**			**11.41**

Avg Wt	$/lb.	avg cost/bird		Avg Wt	$/lb.	avg cost/bird	
5.15	3.95	$ 20.34		5.15	3.95	$ 20.34	

Margin	% Costs/Price			Margin	% Costs/Price		
9.85	51.59%			8.93	56.10%		

Summary of impact		Per Bird	Per Year				
Better feed conversion		$0.76	$1,516.90				
Lower mortality		$0.33	$663.54				
Cost of NutriBalancer		-$0.17	$340.00				
Difference to our bottom line		$0.92	$1,832.10				

Appendix B-1: No-Soy Ration Research

Dickinson College Farm and Fertrell Company Broiler Chicken Research
Author: Alex Smith
Co-Authors: Jenn Halpin and Jeff Mattocks

Abstract
This experiment examined two populations of K-22 broiler chickens. The two populations were kept in nearly identical conditions but were fed two different types of feed. One feed had a base of roasted soybeans, the other had a base of field peas. Data was collected on the feed consumed and weights of the birds. It appears that using a feed based in soybeans can produce a larger and more robust bird.

Introduction
The purpose of this experiment was to observe whether there was a difference between K-22 broiler chickens that were raised under nearly identical conditions and circumstances, but with two different types of feed. The feeds differ in content significantly but perhaps the most notable difference is the use of roasted soybeans in the "Soy feed" (referred to as "soy-based" or "soy") and the use of field peas in the "Non-soy feed" (referred to as "non-soy" or "soy-free"). The ingredients are listed in Tables 1 and 2 as a percent of total feed.

Table 1: Soy-free Feed Ingredients as Percentage

Ingredients	% of Total
Corn Grain Shell	23.25%
Crab Meal	7.50%
Fish Meal, 64%	6.25%
Flax Seed	5.00%
Peas	40.00%
Poultry Nutri-Balancer	3.00%
Wheat	15.00%

Table 2: Soy Feed Ingredients as Percentage

Ingredients	% of Total
Alfalfa Hay	5.00%
Aragonite	1.25%
Corn Grain Shell, ground	50.75%
Fish Meal, Menhaden, 60+%	3.75%
Poultry Nutri-Balancer	3.00%
Soybeans, roasted	31.25%
Spelt	5.00%

Feeding Pasture-Raised Poultry

There exists a debate within the pasture poultry community regarding the use of soy in raising broiler chickens. While the authors do not take sides in the debate, this experiment is meant to add depth to the debate through its findings.

Methods and Materials
The two populations were kept in nearly identical conditions and circumstances. However, they were fed two different kinds of feed from start to finish – no chick starter feed was used.

The birds were started in brooders inside a barn and kept at temperatures as recommended by Joel Salatin in "Pastured Poultry for Profits." At 21 days old they were moved outside into portable chicken tractors. Although the two houses were different designs, the square footage of grass space was identical. When the chickens were moved outside, the houses were moved to fresh pasture daily from 22 days old to 48 days old. From 49 to 65 days old the houses were moved twice daily due to the increasing size, waste load, and demand of fresh pasture of the birds. From 66 days old to slaughter at 81 days the birds had access to the area outside of their respective houses and were each contained by a single, 164-foot feather-net fence. The fence was arranged in such a fashion that each set had identical access to the outside.

The feed and water were also dispensed using the same methods. A single hanging feeder was used in each set of broilers until dietary demands increased. At 49 days old a second feeder was added for both sets. A single five-gallon water dispenser was used in both sets throughout the course of development. The water was filled as it approached empty and thus was never left empty. During the latter half of the birds' lives, the water was refilled everyday. Both the water dispensers and feeders were elevated to chest height for the birds at each stage of their growth.

Data was recorded daily on how much feed and water was provided for the birds as well as how much feed and water were consumed. After slaughter this data was organized to note the feed consumed over the lifetime of the chickens as well as the average feed consumed per bird. In addition to the feed data, sample weights were taken and extrapolated to represent the respective populations. A sample size of 2 to 5 birds was taken 8 times throughout the course of the birds' development at days 1, 8, 12, 21, 37, 45, 64 and 81. This data was used to examine growth patterns and discrepancies between the two populations.

Appendix B-1: No-Soy Research

Results

Figure 1 represents the total feed consumed daily in both populations.

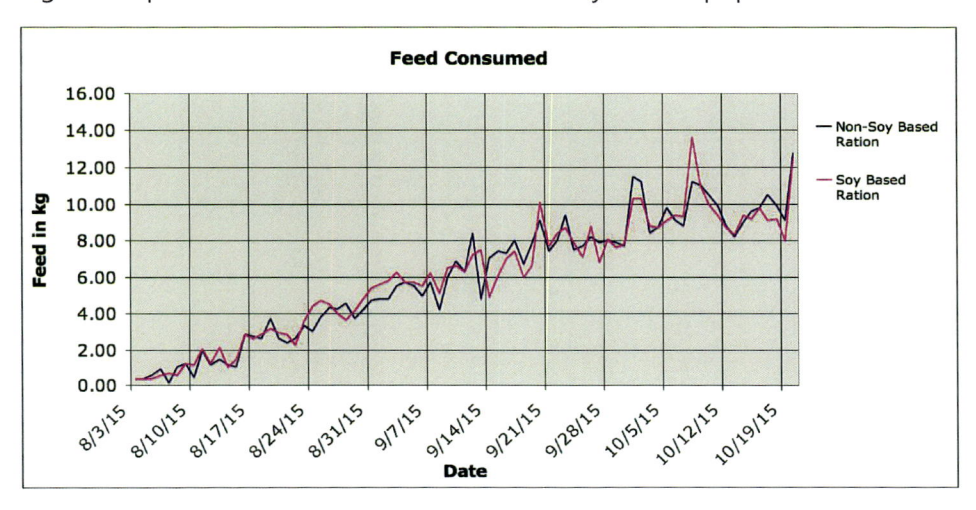

Figure 1: Feed Consumed in Kilograms

Figure 2 reflects the average feed consumed per bird. Mortality and population size were factored into the calculations.

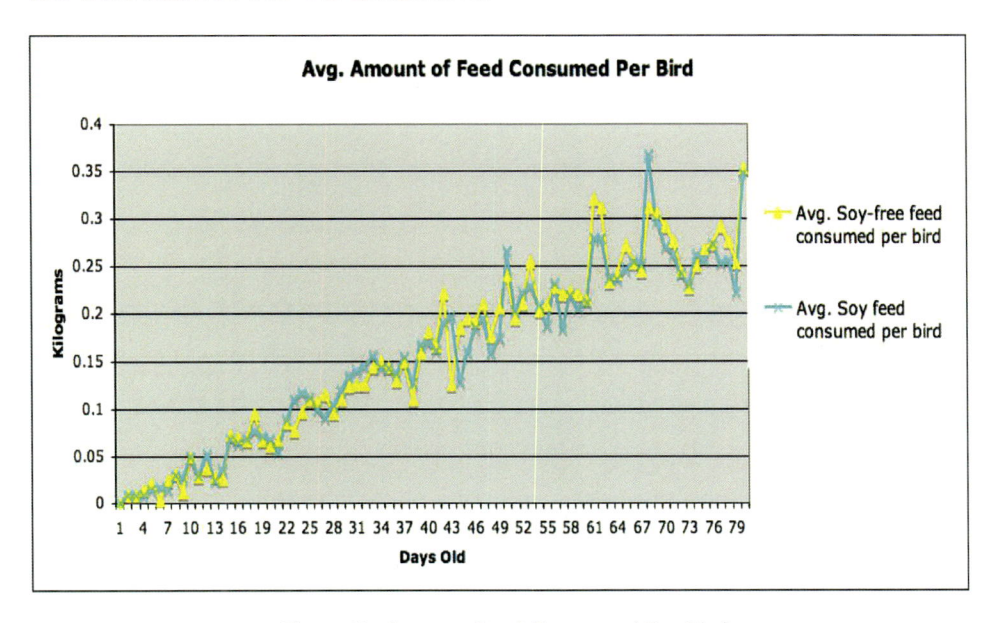

Figure 2: Average Feed Consumed Per Bird

Feeding Pasture-Raised Poultry

Figure 3 and Table 3 represent the results of the sample weights.

Date	Avg. Live Weight of Soy Birds (kgs)	Avg. Live Weight of Soy-free Birds (kgs)	Sample size
1	0.03	0.03	5
8	0.09	0.07	5
12	0.23	0.19	3
21	0.40	0.28	2
37	1.2	1.03	4
45	1.7	1.50	4
64	1.85	1.55	4
81	3.52	3.28	5

Table 3: Average Live Weight Per Population

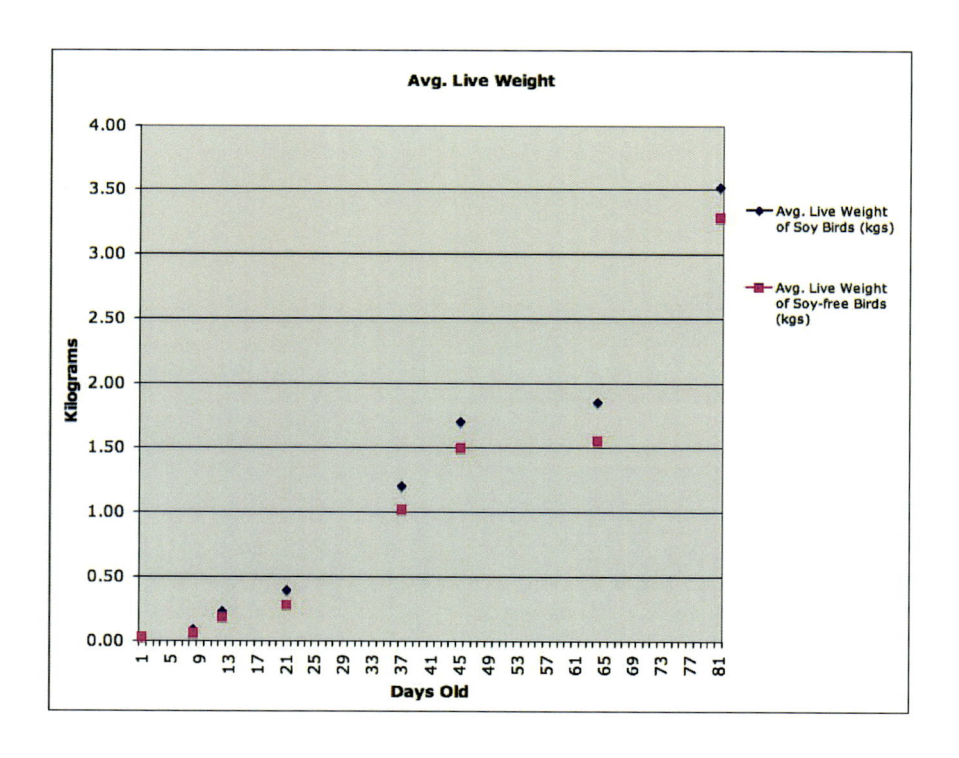

Figure 3: Average Live Weight Per Population

Appendix B-1: No-Soy Research

Figure 4 was created by finding the percent of body weight that was consumed in feed per bird throughout the course of the birds' development.

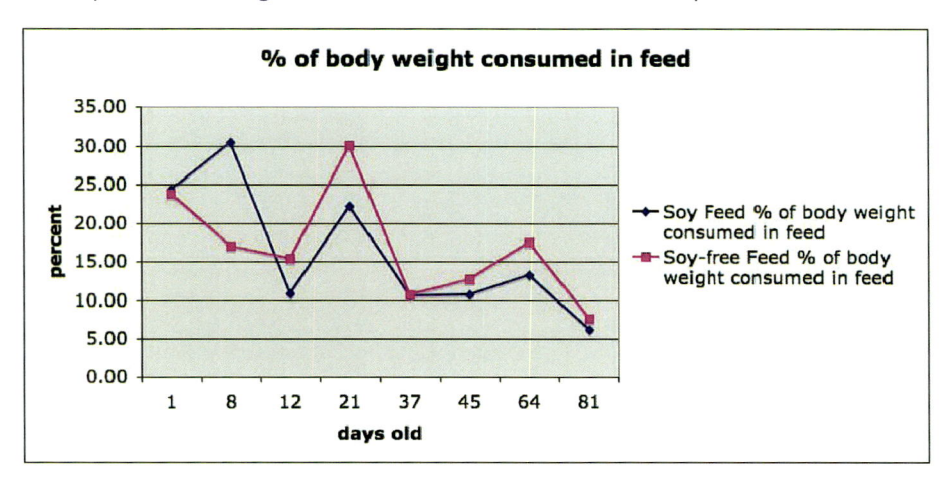

Figure 4: Percent of Body Weight Consumed In Feed

Observations

The birds raised on soy-based feed weighed an average of 2.24 kilograms after being slaughtered and processed. The birds raised on the soy-free feed weighed an average of 2.07 kilograms after being slaughtered and processed. Respectively this was 63.6% and 63.1% of the average live weights on the day of slaughter. It was observed by those participating in the experiment that the combs and wattles appeared larger, more full, and more red in the birds fed the soy-based ration. The shanks on those birds also appeared more yellow and shiny. Finally, the feathers also appeared more full and red. The temperament of the two populations seemed indistinguishable until the day of slaughter when the birds fed the soy-free ration seemed to be more resistant to being collected for slaughter. The worker responsible for collecting the birds reported that the birds fed the non-soy ration were not only more resistant to capture but more aggressive in that they were pecking his hands. It must be noted that the birds raised on the soy-free ration were collected after the other population. Perhaps the disposition is attributable to a higher stress level at time of collection or changes in daylight.

The mortality rates for the two populations seemed relatively normal and predictable. Six birds died from the population fed the soy-free ration and five died from the other population. One mortality from each set was due to an accidental death while moving the house. It should also be noted that a bird from the population fed the soy-based ration appeared particularly lethargic and ill on the 70th day. This bird was prematurely, but intentionally, slaughtered and processed on that afternoon. No autopsies were performed on the other birds so there is no known cause of death.

Feeding Pasture-Raised Poultry

Analysis

Each population consumed significantly similar amounts of feed per bird and they were treated nearly identically. This, in conjunction with the representative weights of the birds (both live and processed) seems to indicate that the birds fed the soy-based feed were more efficient at utilizing feed for growth.

However, this is not necessarily suggesting that the feed consumed was the direct factor responsible for the weight gain. Among other discrepancies, there exists variation in caloric values, protein content, fat content, and moisture content in the feed. The weight difference could be attributable to a variation in overall health, vigor or grazing ability that arose from the dissimilarity in the feeds. The decrease in feed consumed per body weight lends credence to the latter theory that the feed isn't directly the cause of a larger bird. Given that growth continues to accelerate but the amount of feed consumed with regards to body weight decreases it seems that birds are more heavily relying on grazing for their growth and body maintenance. Regardless, the data suggests that distributing the soy-based feed produces, on average, a larger bird. Furthermore, given the bird quality observations, it appears that the birds fed the soy-based feed were overall more robust and healthy. It should also be noted that the ratios of dressed weight to live weight were remarkably similar in each population. Thus neither population appeared more efficient at producing meat per bird.

While the following observation is beyond the scope of the project, it ought to be noted that there appears to be a peculiarity regarding the growth data. The average live weight data generally seems to proceed linearly throughout the course of the birds' development. However, further inspection suggests that from days 21 to 37 and days 64 to 81 growth appears to accelerate. Day 21 was the day that the birds were moved outside. Day 64 was when the birds were given access to the area outside of their house and were contained by the feather-net fencing. These apparent spikes are perhaps a result of a dramatic increase to available grazing and movement area.

Quality Control

While the experiment was fairly well controlled there were several quality control issues that could have possibly impacted results. It must be noted that the Dickinson College Farm is not a research facility – it is an operating and producing farm. Furthermore, its employees, both managers and interns are not researchers, they are farmers or farmers-in-training.

Throughout the course of the experiment there were three people who were responsible for the daily care of the birds and recording data. Given that the amount of feed to dispense was fairly subjective, this could have played a role in the results. Also, given the diversity of responsibilities beyond the broiler experiment, farm employees and managers were unable to provide ideal consistency of treatment. This applies to timing of feeding and moving each day. Furthermore,

Appendix B-1: No-Soy Research

the frequency and sample size of weights was unfortunately low. More frequent and larger samples could have yielded more reliable and significant data. Again, this was limited by the time constraints of the experimenters.

The two house models, while designed to provide similar access to fresh pasture, had significant differences that could have affected data. The birds fed the soy-free ration were kept in a hoop house design and the birds fed the soy ration were kept in a Salatin-style pen. Jeff Mattocks noted that the hoop-house design is more biologically suitable for the health of the birds. Thus the birds fed the soy-free ration were more predisposed to weight gain.

Despite these issues, pending statistical analysis of the results will either strengthen or weaken the significance of the recorded data. Though the numbers appear significant and consistent.

Acknowledgements
Special thanks are due to Matt Steiman for his participation in the project – serving as the nighttime caretaker of the chickens and providing insight and prospective to the analysis of the data. Thanks to Susan Beal for her observations. Thanks to Katelyn Repash, Claire Fox and the Dickinson College Farm student worker crew for their contributions as well.

References For This Research
Dudley-Cash, William A. 1994. *Feedstuffs*. October 3.
Dudley-Cash, William A. 1998. *Feedstuffs*. April 6.
Ferket, Peter R. 2000. Feedstuffs. September 4
Ewing, W. Ray. 1963. *Poultry Nutrition*. 5th edition. The Ray Ewing Company, Pasadena, CA
Leeson, S. and J.D. Summers, 1997. *Commercial Poultry Nutrition*. Second Edition. Available from: University Books, P.O. Box 1326, Guelph, Ontario, Canada, N1H 6N8.
Scott, Milton L., Malden C. Nesheim, Robert J. Young 1982. *Nutrition of the Chicken*, 3rd Edition. Cornell University, Ithaca, NY.
Morrison, Frank B. 1951. *Feeds and Feeding*. 21st edition. Morrison Publishing Company, Ithaca, NY.
Robinette, Jack A. 2002. MBA, Advanced Bio-Chemistry (Nutritional Consultant), Hershey, PA (retired). Personal communication.

Feeding Pasture-Raised Poultry

Appendix B-2: 2012 No-Soy Ration Research

Continued Soy and No-Soy Ration Research

Soy and no-soy feed comparison research was continued in 2012 on Dickinson College Farm with K-22 Red Broilers, at the Rodale Institute with Cornish cross birds, and at Jason and Heather Fritz's Farm with Bard Silver Cockerels.

On each farm a single breed was raised, split into two groups. Otherwise treated exactly the same, the difference between the groups was that one group was fed a soy ration and one a no-soy ration.

Table 1: Soy-free Feed Ingredients

Ingredients	Pounds
Corn Grain Shell	465
Crab Meal	150
Fish Meal, 64%	125
Flax Seed	100
Peas	800
Poultry Nutri-Balancer	60
Wheat	300
Total	2000

Nutrients:
Crude Protein	19.7%
Crude Fat	4.4%
Crude Fiber	4.6%
Energy	1,300 Kcal/lb

Table 2: Soy Feed Ingredients

Ingredients	Pounds
Alfalfa Hay	100
Aragonite	25
Corn Grain Shell, ground	1015
Fish Meal, Menhaden, 60+%	75
Poultry Nutri-Balancer	60
Soybeans, roasted	625
Spelt	100
Total	2000

Nutrients:
Crude Protein	19.7%
Crude Fat	8.1%
Crude Fiber	4.5%
Energy	1,379 Kcal/lb

Appendix B-2: 2012 No-Soy Research

Results: Dickinson College Farm
- Received 82 day-old K-22 Cockerel chicks (1 dead upon delivery) July 27. 2012.
- Randomly split into two groups, soy (n=40) and non-soy (n=41).
- Two pens of equal ground area (approx. 64 ft^2) but different overall design were chosen, one for each group.
- Both groups of chickens were housed entirely in pens until the week of 9/17, when they were allowed access to an area enclosed with electronetting.
- Pens were closed nightly after this period, and opened around 8 AM each morning.
- Food was initially provided ad libitum from a hanging feeder; for a brief period from 9/9 to 9/14, occasionally one or both groups would consume all the food provided to them in less than 24 hours, until a second feeder was added to both pens 9/15.
- Water was provided ad libitum at all times through a 5- or 3-gallon waterer in each pen.

Dickinson				
Soy Fed Group		Total Units	Soy Free Group	Total Units
Total feed Consumed		579.48 lbs	Total consumed	684.4 lbs
Birds in Group		37 Each	Birds in Group	35 each
Feed Consumed per bird		15.66 lbs	Feed Consumed per bird	19.55 lbs
Average Live Weight		6.577 lbs	Average Live Weight	6.491 lbs
Average Carcass Weight		4.76 lbs	Average Carcass Weight	4.457 lbs
FCR Live weight		2.38 lbs	FCR Live weight	3.013 lbs
FCR Carcass Weight		3.29 lbs	FCR Carcass Weight	4.387 lbs
Feed cost/Lb		0.406 lbs	Feed Cost/Lb	0.475 lbs
Total Feed Cost		$ 6.36	Total Feed Cost	$9.29
Cost per Lb Carcass		$ 1.34	Cost per Lb of Carcass	$2.08 lbs

Below: Dickinson shelter styles: Flat top and hooped

Feeding Pasture-Raised Poultry

Dickinson College Farm Continued

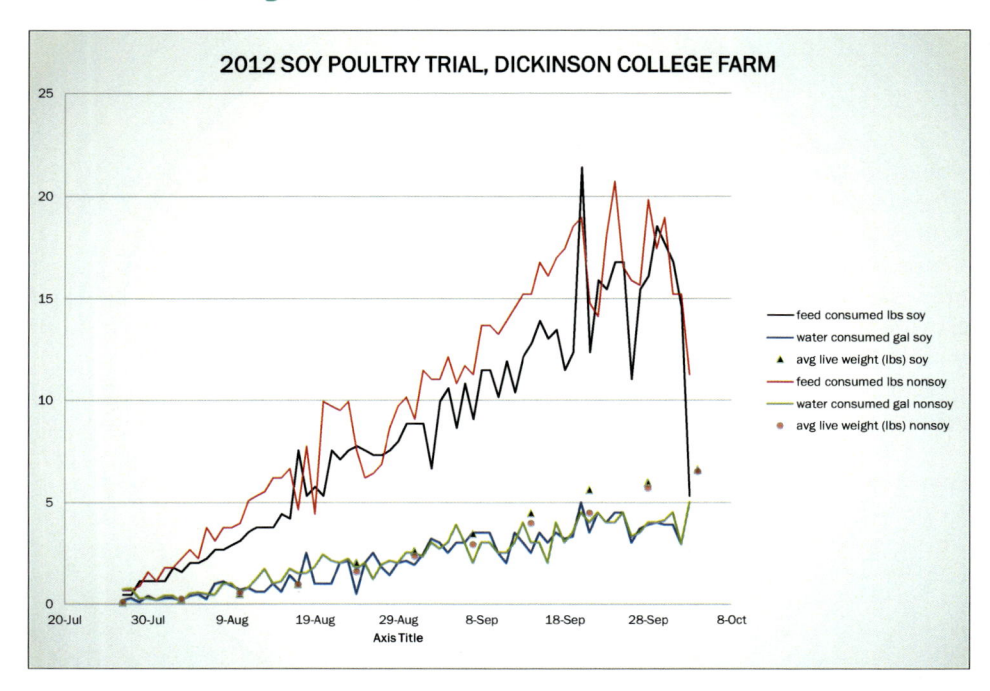

2012 SOY POULTRY TRIAL, DICKINSON COLLEGE FARM

Phytoestrogen Analysis

ug/100 g Sample Name	wet weight Daidzein	Genistein	Glycitein + Biochanin
DSF 1			24.3
DSF 2			2.8
DSF 3	2.3		2.7
DSF 4	3.4		7.5
DNS 1			2.9
DNS 2			
DNS 3			
DNS 4			

ug/100 g Sample Name	wet weight Formononetin	Coumestrol	Apigenin	Total Isoflavones:
DSF 1				24.3
DSF 2		1.5		2.8
DSF 3				4.9
DSF 4		2.1		10.9
DNS 1				2.9
DNS 2				
DNS 3				
DNS 4				

Appendix B-2: 2012 No-Soy Research

Results: Rodale Institute
• This experiment examined two populations of 50 Cornish cross broiler chickens.
• The two populations were kept in nearly identical conditions but were fed two different types of feed.
• One feed had a base of roasted soybeans, the other had a base of field peas.
• Data was collected on the amount of feed consumed and on the weight gains of the birds. It appears that using a soybean-based feed can produce a larger and more robust bird

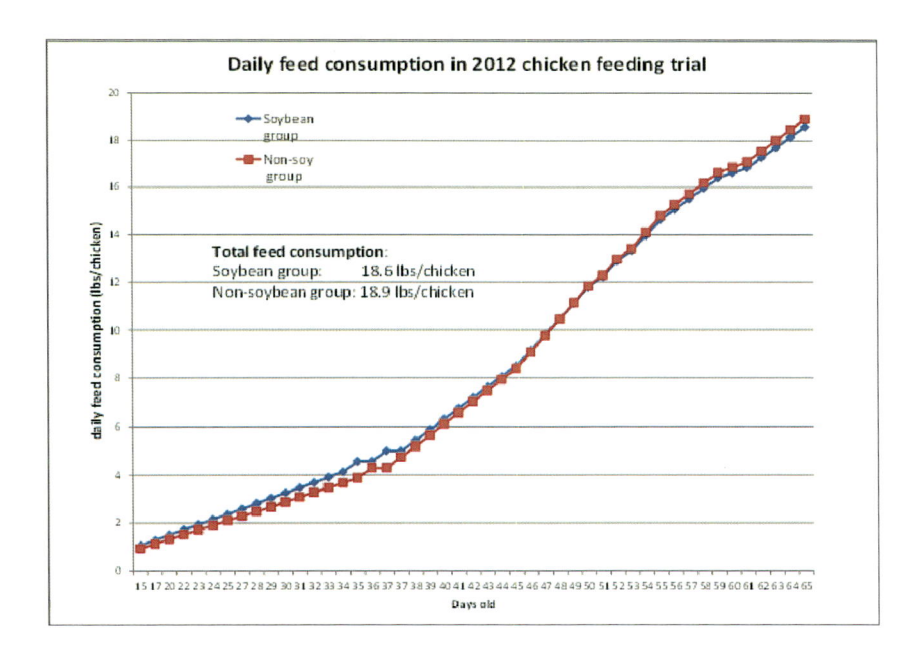

Daily feed consumption in 2012 chicken feeding trial

Total feed consumption:
Soybean group: 18.6 lbs/chicken
Non-soybean group: 18.9 lbs/chicken

Rodale Shelters

Feeding Pasture-Raised Poultry

Rodale Institute Continued

Weekly weight gains for each group (based on a sample size of 5 per group)

Date	Days old	Soy group (g/bird)	Non-soy group (g/bird)	Soy group (lbs/bird)	Non-soy group (lbs/bird)	weight gain difference (%)
09/26/12	15	171	170	0.4	0.4	1
10/03/12	22	386	329	0.8	0.7	17
10/10/12	29	777	615	1.7	1.4	26
10/17/12	36	1,254	667	2.8	1.5	88
10/24/12	43	1,402	874	3.1	1.9	60
10/31/12	50	2,121	1,572	4.7	3.5	35
11/08/12	58	2,319	1,976	5.1	4.3	17
11/15/12	65	3,034	2,466	6.7	5.4	23

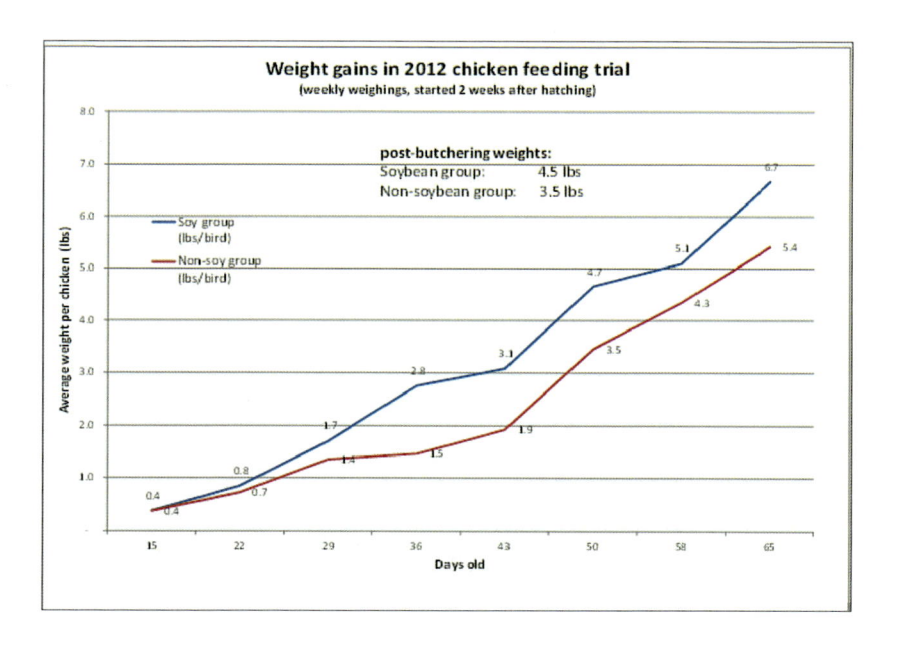

Weight gains in 2012 chicken feeding trial
(weekly weighings, started 2 weeks after hatching)

post-butchering weights:
Soybean group: 4.5 lbs
Non-soybean group: 3.5 lbs

Appendix B-2: 2012 No-Soy Research

Rodale Institute Continued

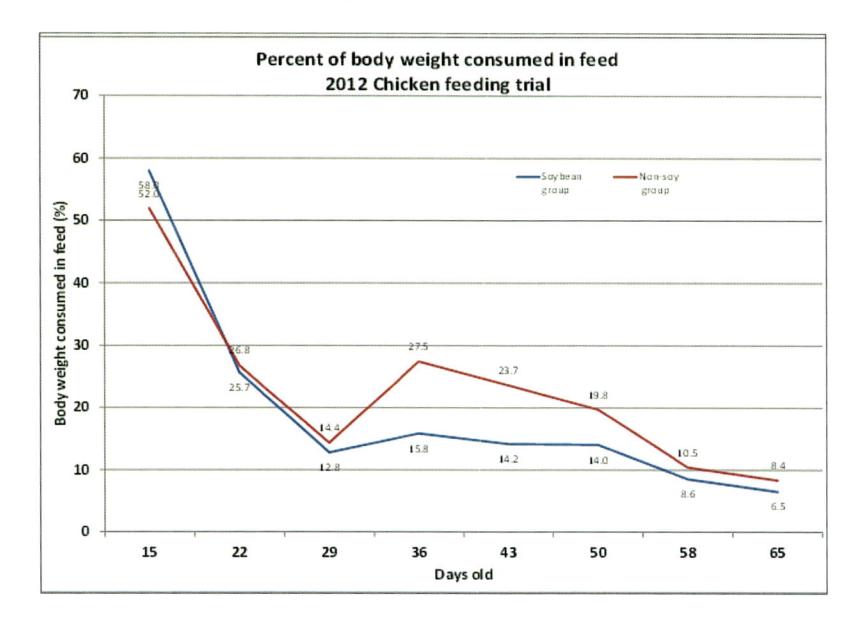

Percent of body weight consumed in feed — 2012 Chicken feeding trial

Rodale			
Soy Fed Group	Total Units	Soy Free Group	Total Units
Total feed Consumed	869.5 lbs	Total consumed	869.4 lbs
Birds in Group	47 Each	Birds in Group	46 each
Feed Consumed per bird	18.50 lbs	Feed Consumed per bird	18.9 lbs
Average Live Weight	6.7 lbs	Average Live Weight	5.4 lbs
Average Carcass Weight	4.5 lbs	Average Carcass Weight	3.5 lbs
FCR Live weight	2.76 lbs	FCR Live weight	3.5 lbs
FCR Carcass Weight	4.11 lbs	FCR Carcass Weight	5.4 lbs
Feed cost/Lb	0.406 lbs	Feed Cost/Lb	0.475 lbs
Total Feed Cost	$ 7.52	Total Feed Cost	$ 8.98
Cost per Lb Carcass	$ 1.67	Cost per Lb of Carcass	$ 2.57 lbs

Feeding Pasture-Raised Poultry

Results: Jason and Heather Fritz

- This experiment consisted of two groups of 50 Bard Silver Cross broiler chickens.
- The two groups were managed in nearly identical ways but were fed two different types of feed.
- One feed had a base of roasted soybeans,
- The other had a base of field peas, linseed, fish and crab meal.
- Data was collected on the amount of feed consumed daily.
- Weights were taken weekly to show gains of the birds.

Fritz			
Soy Fed Group	Total Units	Soy Free Group	Total Units
Total feed Consumed	1226 lbs	Total consumed	1290 lbs
Birds in Group	46 Each	Birds in Group	47 each
Feed Consumed per bird	26.65 lbs	Feed Consumed per bird	27.45 lbs
Average Live Weight	7.275 lbs	Average Live Weight	6.778 lbs
Average Carcass Weight	4.665 lbs	Average Carcass Weight	4.47 lbs
FCR Live weight	3.66 lbs	FCR Live weight	4.05 lbs
FCR Carcass Weight	5.71 lbs	FCR Carcass Weight	6.14 lbs
Feed cost/Lb	0.406 lbs	Feed Cost/Lb	0.475 lbs
Total Feed Cost	$ 10.83	Total Feed Cost	$ 13.04
Cost per Lb Carcass	$ 2.32	Cost per Lb of Carcass	$ 2.92 lbs

Appendix B-2: 2012 No-Soy Research

Fritz Farm Continued

Phytoestrogen Analysis

ug/100 g Sample Name	wet weight Daidzein	Genistein	Glycitein + Biochanin
FSF 1			40.7
FSF 2		5.3	11.0
FSF 3			6.6
FSF 4		3.9	3.0
FNS 1		12.3	37.7
FNS 2			18.2
FNS 3			10.7
FNS 4			8.7

Sample Name	wet weight Formononetin	Coumestrol	Apigenin	Total Isoflavones:
FSF 1				40.7
FSF 2				16.3
FSF 3				6.6
FSF 4				6.9
FNS 1	8.6	14.2	29.2	50.1
FNS 2	3.8			18.2
FNS 3	4.3	19.3		10.7
FNS 4				8.7

Fritz Field Shelters

Feeding Pasture-Raised Poultry

Appendix C-1: Feed Ingredient Values and Spreadsheet Ration Calculator

Ingredients:	LBS	%/Lb %CP	Lb/Mix Protein	%/Lb Fat	Lbs/Mix Fat	%/Lb Fiber	Lbs/Mix Fiber	Kcal/Lb Energy	Kcal/Mix Energy
Alfalfa Meal		17%		3.0%		24.0%		640	
Aragonite		0.00%		0.0%		0.0%			
Barley		12%		2.0%		5.0%		1250	
Corn Gluten Meal		60.0%		2.0%		5.0%		1700	
Corn Grain Shell		8%		3.5%		2.9%		1520	
Crab Meal		32.0%		2.0%		6.0%		675	
Dicalcium Phosphate		0.00%		0.0%		0.0%			
DL Methionine		99.00%		0.0%		0.0%			
Fish Meal, 60%		60%		7.7%		3.5%		1300	
Fish meal, Sea-Lac		63%		10.0%		1.0%		1300	
Hulless Oats		13.0%		4.0%		10.0%		1100	
Hulless Oats		16.0%		6.0%		10.0%		1100	
Lysine		99.00%		0.0%		0.0%			
Oats		10.0%		4.0%		3.0%		1100	
Oil, Canola		2%		99.0%		5.0%		3900	
Oil, Coconut		2%		99.0%		5.0%		3900	
Oil, Soy Bean		0.0%		4.0%		10.5%		3950	
Peas		22.4%		4.0%		6.3%		1300	
Poultry Nutri-Balancer		0.00%		0.4%		1.2%		115	
Shell Corn Grain		8%		3.5%		2.9%		1500	
Soy Bean Meal, 48%		47.0%		3.5%		5.0%		1020	
Soybean Meal, Exp.		44.0%		7.0%		5.0%		1300	
Soybeans, Roasted		38%		18.0%		5.0%		1500	
Sunflower, Black Oil		18%		27.7%		31.0%		1200	
Triticale		11%		1.5%		2.9%		1430	
Wheat		12.0%		6.0%		5.0%		1380	
Wheat		9%		3.5%		5.0%		1400	
Wheat		14%		1.9%		2.9%		1440	
Whole Canola		22%		40.0%		5.0%		2150	
Total									

Directions for use of Spreadsheet Ration Calculator:
1. Columns provided between each nutritional component are for calculations.
2. Multiply the desired pounds of each feedstuff with corresponding component value. Place value in the right side column.
3. When all of the feedstuffs have been calculated, add each column of calculated values vertically downward to arrive with a total at the bottom of each column in the totals row.

Appendix C: Feed Values & Ration Calculator

Appendix C-2: Feed Ingredient Values and Spreadsheet Ration Calculator

Ingredients:	%/Lb Calcium	Lbs/Mix Calcium	%/Lb Phos	Lbs/Mix Phos	%/Lb Salt	Lbs/Mix Salt	%/Lb Sodium	Lbs/Mix Sodium
Alfalfa Meal	3.0%		0.25%					
Aragonite	37.0%		1.0%		0%		0.10%	
Barley	0.1%		0.40%		0%		0.02%	
Corn Gluten Meal	0.2%		0.58%		0%		0.02%	
Corn Grain Shell	0.0%		0.25%		0%		0.02%	
Crab Meal	15.0%		1.70%		2%		0.85%	
Dicalcium Phosphate	22.0%		18.5%				0.10%	
DL Methionine	0.0%		0.0%					
Fish Meal, 60%	3.0%		2.00%		1%		0.40%	
Fish meal, Sea-Lac	5.0%		3.00%		1%		0.34%	
Hulless Oats	0.1%		0.35%				0.02%	
Hulless Oats	0.1%		0.35%				0.02%	
Lysine	0.0%		0.0%					
Oats	0.1%		0.36%		0%		0.06%	
Oil, Canola	0.0%		0.42%		0%		0.03%	
Oil, Coconut	0.0%		0.42%		0%		0.03%	
Oil, Soy Bean	0.1%		0.30%		0%		0.06%	
Peas	0.1%		0.45%				0.01%	
Poultry Nutri-Balancer	13.9%		10.7%		10.00%		4.60%	
Shell Corn Grain	0.0%		0.25%		0%		0.02%	
Soy Bean Meal, 48%	0.2%		0.58%		0%		0.02%	
Soybean Meal, Exp.	0.2%		0.60%		0%		0.02%	
Soybeans, Roasted	0.3%		0.60%				0.02%	
Sunflower, Black Oil	0.2%		0.56%				0.09%	
Triticale	0.5%		0.30%		0%		0.20%	
Wheat	0.1%		0.40%		0%		0.06%	
Wheat	0.3%		0.58%		0%		0.02%	
Wheat	0.5%		0.41%		0%		0.02%	
Whole Canola	0.4%		0.64%		0%		0.02%	
Total								

4. Divide each total value by the total weight of the mix to determine the component level of the proposed ration.
5. Compare total values with required or desired values and make adjustments as needed.
6. Make copies of spreadsheets prior to making calculations and markings on the original spreadsheet.

Continued...

Feeding Pasture-Raised Poultry

Appendix C-3: Feed Ingredient Values and Spreadsheet Ration Calculator

Ingredients:	UI/Lb Vit A	UI/Mix Vit A	UI/Lb Vit D	UI/Mix Vit D	UI/Lb VitE	UI/Lb Vit E	MG/LB Choline	MG/Mix CHO
Alfalfa Meal							486	
Aragonite					0			
Barley	0							
Corn Gluten Meal							6069	
Corn Grain Shell	750						1100	
Crab Meal							200	
Dicalcium Phosphate								
DL Methionine	0		0					
Fish Meal,60%								
Fish meal, Sea-Lac					0		18955	
Hulless Oats								
Hulless Oats							1400	
Lysine	0		0					
Oats	0							
Oil, Canola	0							
Oil, Coconut	0						486	
Oil, Soy Bean							467	
Peas							467	
Poultry Nutri-Balancer	148400		53600		1667		920	
Shell Corn Grain	750							
Soy Bean Meal,48%							6069	
Soybean Meal, Exp.	0						200	
Soybeans, Roasted							1311	
Sunflower, Black Oil							1364	
Triticale	0						1182	
Wheat	123				0		227	
Wheat							209	
Wheat	0						689	
Whole Canola	0						1311	
Total							200	

Continued...

Appendix C: Feed Values & Ration Calculator

Appendix C-4: Feed Ingredient Values and Spreadsheet Ration Calculator

Ingredients:	ppm/Lb Manganese	MG/Mix Manganese	ppm/Lb Zinc	MG/Mix Zinc	ppm/Lb Copper	MCG/Mix Copper	PPM Selenium	PPM/Mix Selenium
Alfalfa Meal	38.2		38		14.08			
Aragonite	286							
Barley								
Corn Gluten Meal	30		11.36		15			
Corn Grain Shell	6		15		3			
Crab Meal								
Di-Calcium Phosphate	700		130		7			
DL Methionine								
Fish Meal, 60%	286							
Fish meal, Sea-Lac	35.6		151		11.4			
Hulless Oats								
Hulless Oats								
Lysine								
Oats	38.2		38		14.08			
Oil, Canola	16		30		8			
Oil, Coconut	16		30		8			
Oil, Soy Bean	128.7		102		0.3			
Peas	15		25		6			
Poultry Nutri-Balancer	3295.6		2413.4		201		10	
Shell Corn Grain	6		15		3			
Soy Bean Meal, 48%	32.3		60		28			
Soybean Meal, Exp.	40		92		8			
Soybeans, Roasted								
Sunflower, Black Oil	16.00		25.00		5.0			
Triticale	0		32		10			
Wheat	28		17		8.2		0	
Wheat	32.3		60		28			
Wheat	40		34		10			
Whole Canola	30.0		11.4		15.0			
Total								

Continued...

Feeding Pasture-Raised Poultry

Appendix C-5: Feed Ingredient Values and Spreadsheet Ration Calculator

Ingredients:	%/Lb LYSINE	Lb/Mix LYSINE	%/Lb METH	Lb/Mix METH	%/Lb Meth/Cystine	Lb/Mix Meth/Cystine	%/Lb Arginine	Lb/Mix Arginine
Alfalfa Meal	0.80%		0.40%		0.18%		0.38%	
Aragonite								
Barley	0.39%		0.18%		0.40%		0.45%	
Canada Poultry VTM			2.47%		2.47%			
Corn Gluten Meal	1.90%		0.88%		1.90%		3.00%	
Corn Grain Shell	0.52%		0.22%		0.20%		0.39%	
Crab Meal	1.70%		1.18%		0.73%		0.93%	
Dicalcium Phosphate								
DL Methionine			99.00%					
Fish Meal,60%	4.60%		1.70%		2.20%		3.10%	
Fish meal, Sea-Lac	4.70%		1.80%		2.40%		3.23%	
Hulless Oats	0.40%		0.20%		0.42%		0.80%	
Hulless Oats	0.40%		0.20%		0.42%		0.80%	
Lysine	56.00%							
Oats	0.64%		0.21%		0.52%		0.80%	
Oil, Canola	0.00%		0.14%		0.34%		0.42%	
Oil, Coconut	0.00%		0.14%		0.34%		0.42%	
Oil, Soy Bean	4.10%		5.20%		2.20%		2.90%	
Peas	1.32%		0.13%		0.37%		1.49%	
Poultry Nutri-Balancer			2.48%		2.48%			
Shell Corn Grain	0.52%		0.22%		0.20%		0.39%	
Soy Bean Meal,48%	2.70%		0.65%		1.50%		3.20%	
Soybean Meal, Exp.	4.20%		0.60%		1.80%		3.50%	
Soybeans, Roasted	2.40%		0.54%		1.10%		2.80%	
Sunflower, Black Oil	0.54%		0.27%		0.50%		1.20%	
Triticale	0.32%		0.26%		0.26%		0.52%	
Wheat	0.30%		0.25%		0.50%		0.75%	
Wheat	0.32%		0.27%		0.14%		0.16%	
Wheat	0.32%		0.22%		0.48%		0.52%	
Whole Canola	2.40%		0.63%		1.00%		2.80%	
Total								

Appendix D: Commercial Broiler-Roaster Nutritional Requirements

	Both Pre-Starter	Broiler Starter #1	Roaster Starter #2	Broiler Grower #1	Roaster Grower #2	Broiler Finisher #1	Roaster Finisher #2
Approximate Protein	23	22	20	20	18	18	16
Metabolizable Energy (Kcal/Lb)	1386	1386	1318	1432	1364	1455	1386
Calcium (%)	1	0.95	0.95	0.92	0.92	0.9	0.9
Available Phosphorus (%)	0.45	0.42	0.42	0.4	0.4	0.38	0.38
Sodium (%)	0.19	0.18	0.18	0.18	0.18	0.18	0.18
Amino Acids (% of Diet)							
Argine	1.4	1.2	1.1	1.05	0.95	0.9	0.85
Lysine	1.35	1.2	1.05	1.1	0.9	0.9	0.8
Methionine	0.52	0.48	0.42	0.44	0.38	0.37	0.36
Methionine/Cystine	0.95	0.82	0.75	0.73	0.65	0.64	0.61
Tryptophan	0.22	0.2	0.18	0.17	0.15	0.14	0.13
Histidine	0.42	0.4	0.35	0.32	0.3	0.28	0.27
Leucine	1.5	1.4	1.2	1.1	1.0	1.0	0.9
Isoleucine	0.85	0.75	0.6	0.55	0.5	0.47	0.45
Phenylalanine	0.8	0.75	0.65	0.6	0.55	0.53	0.5
Phenylalanine+Tyrosine	1.5	1.4	1.2	1.1	1.0	1.0	0.9
Threonine	0.75	0.7	0.62	0.6	0.55	0.55	0.5
Valine	0.9	0.8	0.7	0.65	0.6	0.58	0.55
Vitamins (per Lb)							
Vitamin A (I.U.)		2955				2659	
Vitamin D (I.U.)		1364				1227	
Choline (mg)		364				327	
Riboflavin (mg)		2.5				2.3	
Pantothenic Acid (mg)		6.4				5.7	
Vitamin B_{12} (mg)		0.006				0.005	
Folic Acid (mg)		0.45				0.41	
Biotin (mg)		0.09				0.08	
Niacin (mg)		18.2				16.4	
Vitamin K (mg)		0.9				0.8	
Vitamin E (I.U.)		13.6				12.3	
Thiamin (mg)		1.8				1.6	
Pyridoxine (mg)		1.8				1.6	

Trace Minerals (PPM)	PPM
Manganese	70
Iron	80
Copper	10
Zinc	80
Selenium	0.3
Iodine	0.4

Note: Trace mineral requirements remain the same for all rations.

The information for Appendices D through K is from *Commercial Poultry Nutrition* by S. Leeson and J.D. Summers. 1997

Feeding Pasture-Raised Poultry

Appendix E: Commercial Layer Nutritional Requirements

Feed Intake	\multicolumn{7}{c}{Laying Hens Feed Intake per Day (Lb)}
	0.28	0.26	0.24	0.22	0.2	0.18	0.16
Approximate Protein (%)	13	14	15.5	17	19	20.5	22.1
Metabolizable Energy (Kcal/Lb)	1227	1227	1275	1295	1295	1295	1318
Calcium (%)	3	3.25	3.5	3.6	3.8	4	4.25
Available Phosphorus (%)	0.35	0.4	0.4	0.42	0.45	0.45	0.47
Sodium (%)	0.17	0.18	0.18	0.19	0.2	0.2	0.22
Amino Acids (% of Diet)							
Argine	0.55	0.6	0.68	0.75	0.82	0.9	0.98
Lysine	0.49	0.56	0.63	0.7	0.77	0.84	0.91
Methionine	0.28	0.31	0.34	0.37	0.41	0.47	0.56
Methionine/Cystine	0.48	0.53	0.58	0.64	0.71	0.8	0.91
Tryptophan	0.1	0.12	0.14	0.15	0.17	0.18	0.2
Histidine	0.13	0.14	0.15	0.17	0.19	0.25	0.25
Leucine	0.64	0.73	0.82	0.91	1	1.09	1.18
Isoleucine	0.43	0.5	0.57	0.63	0.69	0.73	0.82
Phenylalanine	0.34	0.38	0.42	0.47	0.52	0.57	0.61
Phenylalanine+Tyrosine	0.55	0.65	0.75	0.83	0.91	0.99	1.08
Threonine	0.43	0.5	0.57	0.63	0.69	0.73	0.82
Valine	0.49	0.56	0.63	0.7	0.77	0.82	0.91

Vitamins (per Lb)	
Vitamin A (I.U.)	3410
Vitamin D (I.U.)	1150
Choline (mg)	550
Riboflavin (mg)	2
Pantothenic Acid (mg)	4.5
Vitamin B_{12} (mg)	0.005
Folic Acid (mg)	0.35
Biotin (mg)	0.07
Niacin (mg)	18
Vitamin K (mg)	1
Vitamin E (I.U.)	12
Thiamin (mg)	1
Pyridoxine (mg)	1.5
Trace Minerals (PPM)	**PPM**
Manganese	70
Iron	80
Copper	8
Zinc	60
Selenium	0.3
Iodine	0.4

Note: Trace mineral requirements remain the same for all rations.

The information for Appendices D through K is from *Commercial Poultry Nutrition* by S. Leeson and J.D. Summers. 1997

Appendix F: Commercial Turkey Nutritional Requirements

	Starter #1	Starter #2	Grower #1	Grower #2	Developer	Finisher
Approximate Protein (%)	28	26	23	21.5	18	16
Metabolizable Energy (Kcal/Lb)	1320	1365	1385	1410	1455	1500
Calcium (%)	1.4	1.3	1.2	1.3	1	0.9
Available Phosphorus (%)	0.7	0.6	0.5	0.6	0.5	0.4
Sodium (%)	0.18	0.18	0.17	0.17	0.17	0.17
Amino Acids (% of Diet)						
Argine	1.6	1.55	1.4	1.25	1.02	0.95
Lysine	1.7	1.6	1.5	1.3	1.15	1
Methionine	0.62	0.55	0.5	0.47	0.42	0.34
Methionine/Cystine	1	0.9	0.8	0.76	0.67	0.58
Tryptophan	0.28	0.26	0.22	0.21	0.18	0.16
Histidine	0.57	0.55	0.48	0.45	0.37	0.32
Leucine	1.95	1.9	1.6	1.45	1.25	1.15
Isoleucine	1.13	1.05	0.9	0.85	0.72	0.65
Phenylalanine	1.03	0.95	0.82	0.79	0.67	0.6
Phenylalanine+Tyrosine	1.8	1.7	1.5	1.4	1.2	1
Threonine	1	0.95	0.85	0.78	0.67	0.61
Valine	1.2	1.15	1	0.9	0.75	0.65
Vitamins (per Lb)						
Vitamin A (I.U.)		4320		3865	7000	3182
Vitamin D (I.U.)		1230		1100	2200	1000
Choline (mg)		865		775	1500	682
Riboflavin (mg)		2.8		2.5	5	2.27
Pantothenic Acid (mg)		7.8		6.8	15	6.82
Vitamin B_{12} (mg)		0.006		0.006	0.012	0.01
Folic Acid (mg)		0.50		0.35	0.5	0.23
Biotin (mg)		0.12		1	0.15	0.07
Niacin (mg)		36.5		32	60	27.3
Vitamin K (mg)		1.0		1	1.5	0.7
Vitamin E (I.U.)		20.0		14	20	9.1
Thiamin (mg)		1.5		1.4	2.5	1.1
Pyridoxine (mg)		2.75		2.3	3	1.4
Trace Minerals (PPM)		PPM		PPM		PPM
Manganese		80		80		80
Iron		110		110		110
Copper		10		10		10
Zinc		80		80		80
Selenium		0.3		0.25		0.0015
Iodine		0.45		0.45		0.45

The information for Appendices D through K is from *Commercial Poultry Nutrition* by S. Leeson and J.D. Summers. 1997

Appendix G: Meat Duck Nutritional Requirements

	Starter #1	Starter #2	Grower #1	Grower #2
Approximate Protein	22	20	18	16
Metabolizable Energy (Kcal/Lb)	1295	1320	1400	1420
Calcium (%)	0.8	0.83	0.76	0.75
Available Phosphorus (%)	0.4	0.42	0.38	0.35
Sodium (%)	0.18	0.18	0.18	0.18
Amino Acids (% of Diet)				
Argine	1.2	1.05	0.94	0.85
Lysine	1.1	0.96	0.86	0.78
Methionine	0.48	0.43	0.39	0.35
Methionine/Cystine	0.82	0.72	0.66	0.6
Tryptophan	0.22	0.18	0.16	0.15
Histidine	0.44	0.37	0.33	0.29
Leucine	1.56	1.28	1.16	1.04
Isoleucine	0.84	0.69	0.63	0.56
Phenylalanine	0.78	0.64	0.58	0.52
Phenylalanine+Tyrosine	1.52	1.24	1.12	1.01
Threonine	0.76	0.62	0.56	0.5
Valine	0.93	0.77	0.69	0.62

Vitamins (per Lb)	
Vitamin A (I.U.)	3650
Vitamin D (I.U.)	1140
Choline (mg)	365
Riboflavin (mg)	1.8
Pantothenic Acid (mg)	5.5
Vitamin B_{12} (mg)	0.005
Folic Acid (mg)	0.23
Biotin (mg)	0.1
Niacin (mg)	28
Vitamin K (mg)	0.7
Vitamin E (I.U.)	10
Thiamin (mg)	1
Pyridoxine (mg)	1.4

Trace Minerals (PPM)	PPM
Manganese	60
Iron	80
Copper	8
Zinc	60
Selenium	0.2
Iodine	0.4

Note: Trace mineral requirements remain the same for all rations.

The information for Appendices D through K is from *Commercial Poultry Nutrition* by S. Leeson and J.D. Summers. 1997

Appendix H: Commercial Broiler Sample Rations

Ingredients:	Starter 1	Starter 2	Grower 1	Grower 2	Grower 3	Finisher 1	Finisher 2
Corn	1137.5	541	1151.4	1212.8	1082.4	1302.8	1012
Wheat	0	400	0	0	0	0	0
Barley	0	400	0	0	350	0	540
Soybean Meal, 48%	700	535	625	610	460	516	340
Meat Meal, 50%	0	0	20	20	0	0	0
Fish Meal, 60%	0	0	40	0	0	0	0
Fat	70	34	90	90	20	94	20
Ground Limestone	34	32	26	30	30	30	30
Calcium Phosphate, 20% P	30	30	20	30	30	30	30
Iodized Salt	6	6	6	6	6	6	6
Vitamin: Mineral Premix[1]	20	20	20	20	20	20	20
Methionine	2.5	2	1.6	1.2	1.6	1.2	2
Calculated Analysis:							
Crude Protein (%)	22.0	22.0	21.8	20	18	18	16.1
Digestible Protein (%)	17.7	17.7	17.7	16.2	14.4	14.2	12.9
Crude Fat (%)	5.9	5.9	7	7	3.4	7.3	3.4
Metabolized Energy (kcal/kg)	1391	1390	1429	1430	1374	1455	1386
Calcium (%)	1.00	1.00	0.98	0.95	0.95	0.94	0.96
Av. Phosphorus (%)	0.42	0.42	0.42	0.42	0.41	0.41	0.41
Sodium (%)	0.17	0.17	0.17	0.17	0.17	0.17	0.18
Methionine (%)	0.48	0.48	0.4	0.4	0.38	0.37	0.37
Methionine & Cystine (%)	0.82	0.82	0.71	0.71	0.65	0.64	0.61
Tryptophan (%)	0.31	0.31	0.28	0.28	0.25	0.25	0.22
Lysine (%)	1.25	1.25	1.1	1.1	0.93	0.96	0.78
Threonine (%)	0.94	0.94	0.86	0.86	0.75	0.78	0.65

[1] Use of additional Choline Chloride if vitamin premix does not contain this vitamin.

The information for Appendices D through K is from *Commercial Poultry Nutrition* by S. Leeson and J.D. Summers, 1997

Feeding Pasture-Raised Poultry

Appendix I: Commercial Roaster Sample Rations

	Grower 1	Grower 2	Finisher 1	Finisher 2
Ingredients:				
Corn	1200	1296	1058	1286
Wheat	198	280	280	280
Barley	0	0	100	0
Soybean Meal, 48%	440	320	380	320
Fat	80	20	100	30
Ground Limestone	30	30	30	30
Calcium Phosphate, 20% P	26	26	26	28
Iodized Salt	6	6	6	6
Vitamin: Mineral Premix[1]	20	20	20	20
Methionine	1.6	1.4	1.2	1
Calculated Analysis:				
Crude Protein (%)	17	15.1	16.1	15.1
Digestible Protein (%)	13.8	12	12.9	11.9
Crude Fat (%)	6.5	3.8	7.4	4.2
Crude Fiber (%)	2.4	2.5	2.7	2.5
Metabolized Energy (kcal/kg)	1443	1400	1457	1410
Calcium (%)	0.9	0.91	0.91	0.92
Av. Phosphorus (%)	0.38	0.38	0.38	0.39
Sodium (%)	0.17	0.17	0.17	0.17
Methionine (%)	0.37	0.34	0.34	0.32
Methionine & Cystine (%)	0.63	0.56	0.56	0.54
Tryptophan (%)	0.87	0.72	0.8	0.71
Lysine (%)	0.23	0.2	0.22	0.2
Threonine (%)	0.72	0.63	0.67	0.63

[1] Use of additional Choline Chloride if vitamin premix does not contain this vitamin

Appendix J: Commercial Layer Sample Rations

Ingredients:	1	2	3
Corn	1192	1305	1313
Wheat	0	0	240
Soybean Meal, 48%	560	468	220
Fat	40	20	20
Ground Limestone	156	156	156
Calcium Phosphate, 20% P	23	23	23
Iodized Salt	7	7	7
Vitamin: Mineral Premix[1]	20	20	20
Methionine	2	1	1
	2000	2000	2000
Calculated Analysis:			
Crude Protein (%)	18.6	16.9	13
Digestible Protein (%)	17	15.4	11.7
Crude Fat (%)	4.4	3.6	4.2
Crude Fiber (%)	2.3	2.3	2.8
Metabolized Energy (kcal/kg)	1300	1295	1260
Calcium (%)	3.3	3.26	3.25
Av. Phosphorus (%)	0.41	0.4	0.4
Sodium (%)	0.19	0.19	0.18
Methionine (%)	0.42	0.34	0.28
Methionine & Cystine (%)	0.7	0.59	0.46
Lysine (%)	1.02	0.88	0.56

[1] Use of additional Choline Chloride if vitamin premix does not contain this vitamin

The information for Appendices D through K is from *Commercial Poultry Nutrition* by S. Leeson and J.D. Summers. 1997

Feeding Pasture-Raised Poultry

Appendix K: Commercial Turkey Sample Rations

	Starter 1	Starter 2	Grower 1	Grower 2	Finisher 1	Finisher 2
Ingredients:						
Corn	900	1000	1154	1234	1358	1482
Soybean Meal, 48%	772	672	530	444	360	292
Meat Meal, 50%	100	100	200	200	100	60
Fish Meal, 60%	100	100	0	0	0	0
Fat	42	60	58	58	110	92
Ground Limestone	20	16	12	14	20	18
Calcium Phosphate, 20% P	40	25	20	24	26	28
Iodized Salt	4	4	4	4	4	4
Vitamin: Mineral Premix1	20	20	20	20	20	20
Methionine	2.6	2	1.4	1.4	0	4
Calculated Analysis:						
Crude Protein (%)	27.9	26.0	23.2	21.5	17	15
Digestible Protein (%)	25.5	23.7	21.2	19.6	15.5	13.7
Crude Fat (%)	4.4	5.5	5.6	5.8	8.5	6.4
Crude Fiber (%)	3.5	3.4	3.4	3.4	3.4	3.4
Metabolized Energy (kcal/lb)	1315	1364	1394	1405	1477	1489
Calcium (%)	1.58	1.34	1.2	1.27	1.07	0.9
Av. Phosphorus (%)	0.85	0.69	0.63	0.66	0.5	0.46
Sodium (%)	0.18	0.18	0.18	0.18	0.17	0.16
Methionine (%)	0.61	0.56	0.5	0.47	0.78	0.69
Methionine & Cystine (%)	1.05	0.96	0.85	0.79	0.64	0.56
Lysine (%)	1.77	1.62	1.41	1.28	0.89	0.75
Tryptophan (%)	0.37	0.34	0.3	0.27	0.22	0.19
Threonine (%)	1.18	1.1	0.99	0.92	0.87	0.64

[1] Use of additional Choline Chloride if vitamin premix does not contain this vitamin

The information for Appendices D through K is from *Commercial Poultry Nutrition* by S. Leeson and J.D. Summers, 1997

Appendix L: Formulating Rations with the Pearson Square

by T.L. Stanton[*]

Quick Facts...

The Pearson square ration formulation procedure is designed for simple rations. In order for the square to work, follow specific directions for its use. Nutrient contents of ingredients and nutrient requirements must be expressed on the same basis (i.e., dry-matter or "as-fed").

The Pearson square or box method of balancing rations is a simple procedure that has been used for many years. It is of greatest value when only two ingredients are to be mixed. In taking a close look at the square, several numbers are in and around the square. Probably one of the more important numbers is the number that appears in the middle of the square. This number represents the nutritional requirement of an animal for a specific nutrient. It may be crude protein or TDN, amino acids, minerals or vitamins.

In order to make the square work consistently, there are three very important considerations:

1. The value in the middle of the square must be intermediate between the two values that are used on the left side of the square. For example, the 14 percent crude protein requirement has to be intermediate between the soybean meal that has 45 percent crude protein or the corn that has 10 percent crude protein. If barley is used that has 12 percent crude protein and corn that has 10 percent crude protein, the square calculation method will not work because the 14 percent is outside the range of the values on the left side of the square.

2. Disregard any negative numbers that are generated on the right side of the square. Be concerned only with the numerical differences between the nutrient requirement and the ingredient nutrient values.

3. Subtract the nutrient value from the nutritional requirement on the diagonal and arrive at a numerical value entitled parts. By summing those parts and dividing by the total, you can determine the percent of the ration that each ingredient should represent in order to provide a specific nutrient level. Always subtract on the diagonal within the square in order to determine parts. Always double check calculations to make sure that you did not have a mathematical error. It also is very important to work on a uniform basis. Use a 100-percent dry-matter basis for nutrient composition of ingredients and requirements and then convert to an as-fed basis after the formulation is calculated. Corn represents (31.0 / 35.00) x 100 of the ration, or 88.57 percent. Soybean meal represents (4.0 / 35.00) x 100 of the ration, or .43 percent.

Feeding Pasture-Raised Poultry

Check of the calculation:
    88.57 lb corn at 10.0% CP = 8.86
    11.43 lb SBM at 45.0% CP = 5.14
    100.00 lb mixture contains = 14.00 lb CP, or 14 percent.

Using More Than Two Ingredients

It is possible to mix more than two ingredients using the Pearson square. For example, to prepare a 15 percent crude protein mixture that consists of a supplement of 60 percent soybean meal (45 percent crude protein) and 40 percent meat and bone scrap (50 percent crude protein), and a grain mixture of 65 percent corn (9 percent crude protein) and 35 percent oats (12 percent crude protein), take the following steps. Since only two components can be used in the Pearson square method, the ingredients are combined first as follows:

60% SBM x 45% crude protein	= 27.0
40% MBS x 50%	= 20.0
Protein in supplement mixture	47.0%
65% corn x 9.0%	= 5.85
35% oats x 12.0%	= 4.20
Protein in grain mix	10.05%
5.0 parts x 60%	= 3.0 parts SBM
5.0 parts x 40%	= 2.0 parts MBS
32.0 parts x 65%	= 20.8 parts corn
32.0 parts x 35%	= 11.2 parts oats
	37.0
(3.0 / 37.0)	= 8.11% SBM
(2.0 / 37.0)	= 5.41% MBS
(20.8 / 37.0)	= 56.21% corn
(11.2 / 37.0)	= 30.27% corn

Check:
8.11 lb SBM at 45% CP	= 3.65 lb
5.41 lb MBS at 50% CP	= 2.70 lb
56.21 lb corn at 9% CP	= 5.06 lb
30.27 lb oats at 12% CP	= 3.64 lb
100.00 lb contains	15.05 lb or 15% CP

Expressing Feed Composition

The crude-protein value of a feed or the percentage of any other component (e.g., calcium or phosphorus) can be expressed several ways. The two most common methods of expression are on an asfed basis or dry-matter basis. Use the following procedure to calculate composition on a dry-matter basis.

Appendix L: Pearson Square

Crude protein value on an as-fed basis divided by dry-matter content of the feed times 100 equals the crude-protein content on a dry-matter basis. If alfalfa hay is used as an example, the crude protein value is 17 percent on an as-fed basis. On a dry-matter basis, the crude protein value of the hay is calculated as follows: 17 / 0.91 (moisture content of 9 percent) times 100 equals 18.7 percent crude protein.

To determine the total digestible nutrient (TDN) content of the above alfalfa on a dry-matter basis, follow the same procedure: 50 percent (TDN value on an as-fed basis) divided by 0.91 (dry-matter content of the feed) times 100 equals 54.9 percent TDN on a dry-matter basis. Likewise, the crude protein content or the TDN value also can be expressed on the basis of any given dry-matter level. For example, if you use a 90-percent dry-matter basis, use the following calculation. Given a TDN value of 76 percent and a dry-matter content of 86 percent (14 percent moisture), what would be the TDN value of this feed on a 90 percent dry-matter basis?

(76 x .90) / .86 = 79.5 percent TDN on a 90 percent dry-matter basis.

Ration Composition Calculations

If you know the dry-matter composition of a specific ration and want to determine what that composition will be on an as-fed basis for mixing, make the calculations shown in Table 1.

Table 1: Converting from dry matter to as-fed.

Feed	Ration dry-matter composition	Ingredient % dry matter	Calculations		Ration as-fed composition
Corn silage	70	35	70/.35 = 200	(200/233) x 100 =	84.84
Alfalfa	30	90	30/.30 = 33 233	(33/233) x 100 =	14.16

Conversely, if you know the "as-fed" composition of the ration and the dry matter of each ingredient, determine the ration dry-matter composition as shown in Table 2.

Table 2: Converting from as-fed to dry matter.

Feed	Ration dry-matter composition	Ingredient % dry matter	Calculations		Ration as-fed composition
Corn silage	65	35	65/.35 = 22.75	(22.75/54.25) x 100 =	41.94
Alfalfa	35	90	35/.90 = 31.50 54.25	(31.50/54.25) x 100 =	58.06

*Colorado State University Cooperative Extension feedlot specialist and professor, animal sciences. Published by Colorado State University Cooperative Extension. 1995-2001. www.ext.colostate.edu.

Endnotes

p 13   1 Leeson,S. and J.D. Summers, 1997.*Nutrition of the Chicken*, 4th Edition,
       2 Dudley-Cash, William A. 1994. *Feedstuffs*.
       3 Ferket, Peter R. 2000. *Feedstuffs*.
       4 Dudley-Cash, William A. 1998. *Feedstuffs*

p 14   5 Robinette, Jack A. 2002. MBA, Advanced Bio-Chemistry (Nutritional Consultant), Hershey, PA (retired). Personal communication.
       6 Leeson and Summers, 1997

p 18   7 Wikipedia
       8 *Hyline Variety Brown, Commercial Management Guide*, 2006-2008

p 19   9 Leeson and Summers, 1997

p 21   10 Leeson and Summers, 1997, page 8
       11 Morrison, Frank B. 1951. *Feeds and Feeding*. 21st edition. Morrison Publishing Company, Ithaca, NY. 1207 p.

p 22   12 *Traditional Feeding of Farm Animals*, by F.W. Woll, Ph.D., copyright 1915
       13 http://www.merckvetmanual.com/mvm/index.jsp?cfile=htm/bc/201200.htm

p 38   14 Scott, Milton L., Malden C. Nesheim, Robert J. Young 1982. *Nutrition of the Chicken*, 3rd Edition. Cornell University, Ithaca, NY. 562 p.

p 41   15 Morrison, 1951, paragraph 1207
       16 Morrison, 1951, paragraph 845

p 42   17 Morrison, 1951, paragraph 846
       18 Ewing, W. Ray. 1963. *Poultry Nutrition*. 5th edition. The Ray Ewing Company, Pasadena, CA 1475 p.

p 43   19 Morrison, 1951, paragraph 1546

p 43   20 Morrison, 1951, paragraph 1500

Resources

Additional Resources

My most often used references for feed ingredient nutritional values are:

• *Feedstuffs, Reference Issue & Buyers Guide*, Circulation Manager Feedstuffs 191 S. Gary Ave., Carol Stream, IL 60188 Copies of the Reference Issues will cost approximately $40.00

• *Commercial Poultry Nutrition*, Leeson and Summers, 1997. Available from: National Research Council of Canada Building M-58, 1200 Montreal Road, Ottawa, Ontario K1A 0R6, Phone:1-877-672-2672 (in Ottawa, please call 1-613-993-9101) www.nrc.ca

There are many different reference books available. Some are recent publications and some are quite old. Each may make good contributions to your education and understanding of poultry nutrition.

Keep in mind that these references are written for poultry raised under controlled circumstances, all of the data will not apply to pasture-raised poultry. I find myself reading old data and then looking at new data for comparisons. When I find the two periods agree on a subject, I feel it is safe to assume this data is constant and true. When I find conflicting data between the old and the new, I use the data cautiously.

Much of the data available must be interpreted and revised to fit the pasture production model.

Index

Symbols

β-Glucanase 36

A

aflatoxin 34
amino acid 15, 35, 36, 39, 43, 44, 46
appetite 20, 26, 30, 37, 44
ascites 12, 19, 69

B

bagged feed 24-26
barley 29, 35-36, 45, 61, 90-94, 101
bedding 8
bentonite 30
bone meal 24, 44
botulism 17, 70
breeder flocks 12, 58
broilers 8, 9, 12, 20, 22, 23, 30, 32, 38, 42, 52, 58, 61, 73, 76, 95, 99
brooder 8, 9, 16, 20
bulk feed 26-26, 28
by-products 41-44, 58-59, 63

C

calcium 23, 41, 42, 62, 67, 91, 95-102, 104
camelina meal 45
camphlobactor 18
canola 14
carcass weights 9, 22
catfish meal 37
Clostridium Botulinum 17
Clostridium perfringens 17
coccidiosis 9, 18-20, 42
contaminated feed 29-30
copper sulfate 18
corn 8, 13, 24, 29, 34, 35, 39, 40, 58-62, 90-94, 103-104
Cornish cross 9, 11, 20, 58, 61
costs of production 7, 31-33
crumbles, feed 24-25
crustacean meal 37
cystine 36

D

digestive tract 17-22, 42
disease-causing organisms 17-18
ducks 7, 10, 13, 61-62

E

E. coli 9, 18, 39, 70
egg size 11, 51, 62
extruded soybeans 39-40

F

feather development 36
feed conversion 22
feed costs 31-33, 73
feeder space 8-9, 16-17, 67
fiber 35, 36, 45, 46, 90-104
fishmeal 14, 44, 61, 90-94
fishy flavor 14, 45
flax 14, 44, 56, 57, 64, 75, 90-94
forage 10, 14, 41-43, 62

G

garlic 20
genetically modified 15, 32, 33
genetic potential 12
gizzard 21-23
grain brokers 24
grist mills 27
grit 13, 21-23, 35, 61- 67

H

heart attacks 12, 19, 69
heat treated soybeans 27
height of feeder 16-17
heirloom chickens 10
hens 7-10, 14, 19, 21, 31, 38, 42, 48, 51, 66-67
heritage breed layers 66-67
heritage breed turkeys 12
hexane 39
high fiber 35
high production breed 66
hydrogen pyroxide 20

I

illness 22, 68-71
immune system 11, 30, 36, 53
insects 17, 28, 43-44, 66

L

lay cycle 31, 66, 67
least-cost feed ingrediets 24
linoleic acid 14
linseed meal 44
low carcass weights 22
lysine 37, 38

M

mash-type diets 12
meat-type protein 37, 44
menhaden 14, 37
methionine 15, 36-39, 44
milk 18, 20, 41, 42, 63, 69-70

Index

mills 25-27, 31
milo 34
mineral content of water 19
molds 29-30, 34
mortality 11-12, 19, 73, 79
multi-species ration 63-65
mycotoxin 30

N

necrotic enteritis 9, 17, 20-22, 42
nitrates 19
nutritional needs 7, 39, 60, 95-98
nutritional supplements 7

O

oats 35-36, 62-64
Omega-3 13-14, 45
organic 15-19, 31-33, 38, 68, 110
orgainc certification 15-16
oxidize 9, 40, 45
oyster shell 21-23, 67

P

parasites 18
particle size 25, 34, 40
pasture 7-11, 15, 17, 20, 23, 29, 42, 43, 46, 76, 89, 107
pathogens 17, 20, 39
Pearson Square 60, 72, 103, 105
peas 36, 56-57, 75
pelleted feed 13, 25
pentosans 35
Poultry Nutri-Balancer 31, 38, 56-64, 72-75

poults 39
preservatives 24
price per egg 31
probiotics 30, 39
productivity 10, 26
profit 66
pullets 12, 42, 48, 66-67

R

rate of gain 11-12, 63
raw soybeans 39
research 56, 73-81
riboflavin 26, 41, 42
roasted soybeans 39-40
Robinette, Jack 25, 38, 68, 81, 106

S

salmonella 18, 20, 39, 70
salt 37, 39
shelf life 24
solvent-extracted soybean meal 40
soybean meal 24, 39-40, 82-86, 95-96
soy-free 44, 56-57, 64, 75, 79, 81
Spreadsheet Ration Calculator 46, 60, 82-86
sprouts 41
storage 26, 28, 46
sunflower meal 45, 64
supplements 15, 30, 43

T

tannin 34, 36
triticale 35, 64
trypsin inhibitor 27, 39-40
turkeys 10-13, 23, 38, 52-55, 64, 97, 104

U

USDA National Organic Program 15

V

vinegar 19, 20, 69

W

waterer space 9, 16
water quality 19
water temperature 9
wheat 13, 15, 22-24, 29, 35, 61-65, 90-104
whey 41
whole grains 12-13, 21

X

xylanase 35